Esprit Basic Research Series

Edited in cooperation with
the European Commission, DG III/F

Editors: P. Aigrain F. Aldana H. G. Danielmeyer
O. Faugeras H. Gallaire R. A. Kowalski J. M. Lehn
G. Levi G. Metakides B. Oakley J. Rasmussen J. Tribolet
D. Tsichritzis R. Van Overstraeten G. Wrixon

Springer
Berlin
Heidelberg
New York
Barcelona
Budapest
Hong Kong
London
Milan
Paris
Santa Clara
Singapore
Tokyo

Walter Maurel   Yin Wu
Nadia Magnenat Thalmann
Daniel Thalmann

Biomechanical Models for Soft Tissue Simulation

With 77 Figures

Springer

Walter Maurel
Prof. Daniel Thalmann

Computer Graphics Lab
Swiss Federal Institute of Technology
CH-1015 Lausanne, Switzerland

Yin Wu
Prof. Nadia Magnenat Thalmann

MIRALab, Department of Information Systems
University of Geneva
24, rue Général Dufour
CH-1211 Geneva 4, Switzerland

Cataloging-in-Publication Data applied for

Die Deutsche Bibliothek – CIP-Einheitsaufnahme

Biomechanical models for soft tissue simulation / W. Maurel ... –
Berlin; Heidelberg; New York; Barcelona; Budapest; Hong Kong; London; Milan;
Paris; Santa Clara; Singapore; Tokyo: Springer, 1998
(ESPRIT basic research series)
ISBN 3-540-63742-7

CR Subject Classification (1991): J.3, I.3.5, I.3.7

ISBN 3-540-63742-7 Springer-Verlag Berlin Heidelberg New York

Publication No. EUR 18155 EN of the European Commission, Dissemination of Scientific and Technical Knowledge Unit, Directorate-General Information Technologies and Industries, and Telecommunications, Luxembourg.

Printed in Germany

Typesetting: Camera-ready by authors
SPIN 10486038 45/3142 - 5 4 3 2 1 0 - Printed on acid-free paper

To Victor

Foreword

Information Technology is having an increasing influence on medicine. This can be readily observed by anybody visiting a hospital or consulting a doctor, or even by going to the chemist. A range of new medical instruments, new scanners, new on-line diagnostics as well as more effective distribution methods all increasingly contain IT elements that enable new more effective medical tools and services.

But there is also a lot going on "behind the scenes", at the research and development level, that will greatly influence the medical tools and services of tomorrow. In this respect, Esprit, the European Commission's IT research program, has been supporting a range of research and development projects that are contributing to the opening of new avenues in the medical field.

At the 1997 European IT Conference, a special exhibit has highlighted Esprit's contributions in the area, including, for example projects that have succeeded in developing:

- a prototype visual prosthesis linked to the optic nerve (MIVIP)
- implantable blood micro-pumps (IMALP)
- new low-power pacemakers and defibrillators (HIPOCRAT)
- new high-quality 3D ultrasound images (NICE)
- realistic computer models of the human musculo-skeletal system (CHARM)

CHARM, the last – but not least – of these, is further elaborated in this book. The contribution of this project has been the creation of a toolkit of synthetic computer models, particularly focusing on the musculo-skeletal system of the upper torso and shoulder. As a flexible tool it has the potential to empower practitioners in the areas of biomedicine, biomechanics and rehabilitation. CHARM involves some of Europe's top researchers in computer graphics and the modeling of deformable objects, as well as medical practitioners, and it will be interesting to see how this toolset evolves in the future.

All the projects mentioned above will probably contribute either directly or indirectly to the evolving medical field, each bearing hope of contributing to the ever increasing medical quality and health standards available to us. You never know – in the future the results of one of these projects might affect many lives – even yours or mine!

Jakub Wejchert
Esprit Long Term Research

Preface

In November 1993, a three-year Basic Research Project was launched by the European Commission within the ESPRIT program under the acronym CHARM (A Comprehensive Human Animation Resource Model, see Appendix). It has involved several teams working in the areas of: computer animation (like the authors), image processing, civil engineering, system control, anatomy, and orthopedics. CHARM has aimed at developing a comprehensive human resource database and a set of software tools allowing the modeling of the complex human musculoskeletal system and the simulation of its dynamics, including the finite element simulation of soft tissue deformation and muscular contraction. Originally, this book was an internal deliverable report aiming at providing an extensive preliminary review of biomechanical models for soft tissues (CHARM D4). Complemented with physiological observations and theoretical mechanics, the report has appeared to the E.U. experts as an original synthesis on biomechanical modeling suitable for a publication. In this scope, the content has been re-organized and extended with some notions of finite element and incremental resolution methods, in order to provide a reference document for performing similar analysis. Literature offers numerous elaborated manuals investigating deeply, but separately, physiology, theoretical mechanics, soft tissues biomechanics, and numerical methods. Very few of them allow a global understanding of these disciplines and their interrelationships in the scope of a biomechanical simulation. It is with this purpose in mind that the present book has finally been composed.

Soft Tissue Physiology. The first chapter collects some physiological descriptions and experimental observations concerning soft tissues, especially tendons, muscles, and skin. It aims at providing a basic knowledge on soft tissue structures and mechanical behaviors.

Theoretical Mechanics. The second chapter provides basic theoretical definitions of nonlinear mechanics and constitutive relationships to help for further understanding of the experimental relationships reviewed in Chapter 4. In particular, it aims at unifying the various definitions for elasticity and viscoelasticity which may be encountered in the literature.

Resolution Methods. The third chapter offers brief presentations of the finite element method and the incremental formalisms usually applied for the resolution of nonlinear problems. Rather than a reference manual, this chapter aims at giving an idea of the way nonlinear mechanical problems may be solved for a simulation.

Constitutive Modeling. The fourth chapter provides a general overview of the available experimental biomechanical models for soft tissues. More than listing them all, its objective is to present the diversity of forms that biomechanical models may take, by the light of the theoretical mechanics presented in Chapter 2.

Muscle Contraction Modeling. The fifth chapter investigates, in particular, the biomechanical modeling of muscle contraction, with a particular interest for global output force models of individual actuators, consistent with the internal contraction force distribution.

Application Perspectives. The sixth chapter proposes some approaches for finite element implementation and dynamic simulation using biomechanical models for muscle, tendon, and skin, such as those presented in Chapters 4 and 5.

More generally, this book has been composed in the scope of a finite element simulation using biomechanical constitutive relationships. It neither extensively develops the different areas concerned nor addresses the technical aspects of a finite element implementation. Our work has been to investigate the theoretical mechanics, review the state of knowledge in biomechanics, propose some elements from these analyses, and make suggestions before practical investment. This book will be useful to other novice research teams beginning with a similar analysis for understanding how to proceed with biomechanical modeling.

CHARM has been supported by the European Commission, and for the Swiss partners, by the Federal Office of Education and Science.

We are grateful to Dr. J. Wejchert, project officer, and Prof. R.A. Earnshaw, project reviewer, as well as to the other partners of the CHARM project: Ecole des Mines de Nantes (EMN), Universitat de Les Illes Balears (UIB), Universität Karlsruhe (UK), and Instituto Superior Técnico (IST) for the instructive collaboration we have benefitted from.

We finally would like to thank, in particular, Véronique Willi for correcting the English, Paolo Baerlocher and Dr. Prem Kalra for reporting mistakes, Luisa Rita Salvado for our fine privileged collaboration, and Professors Martins and Pires for their general advice about the composition.

July 1997

W. Maurel
Y. Wu
N. Magnenat Thalmann
D. Thalmann

Table of Contents

Nomenclature

$\mathbf{A}_n$	effective stiffness matrix at step n	(3.82)	72
$a(P),\ s(P)$	fiber activation, strength distribution functions	(5.41)	136
$\mathbf{B} = \left[B_{ij}\right]$	left Cauchy–Green dilation tensor	(2.12)	28
$\mathbf{B} \equiv \mathbf{B}(\mathbf{U})$	strain-displacement differential operator	(3.12)	56
$\mathbf{B}_n^L$	linear strain- displacement operator	(3.45)	64
$\mathbf{B}_n^{NL}$	nonlinear strain-displacement operator	(3.45)	64
$\beta_i,\ \gamma_i$	bulk and shear relaxation times	(2.116)	47
$\mathbf{C} = \left[C_{ij}\right]$	right Cauchy–Green dilation tensor	(2.23)	29
$\mathbf{c} = \left[c_{ij}\right]$	Cauchy strain tensor	(2.13)	28
$\mathbf{D} = \left[D_{ij}\right]$	damping matrix	(3.19)	57
$\mathbf{D}_n$	tangent damping matrix	(3.76)	71
$\tilde{\mathbf{D}}_n$	modal tangent damping matrix	(3.91)	75
$\mathbf{D}^P,\ \mathbf{D}^Q$	bulk and shear relaxation matrix	(2.115)	46
$\mathbf{D}^V$	material damping matrix for linear viscoelasticity	(2.103)	44
$\mathbf{D}^{\delta}$	tensorial shape function for fiber direction $\boldsymbol{\delta}$	(6.70)	164
$\mathbf{d}(P)$	tangent vector in P to the muscle fiber	(5.34)	133
$\boldsymbol{\delta}(P)$	unitary contraction stress vector direction	(5.41)	136
$\boldsymbol{\partial}_n$	fiber direction at n in Lagrangian configuration	(6.68)	163
$\boldsymbol{\delta}_n$	nodal fiber direction vector at n	(6.70)	164
$\mathbf{E} = \left[E_{ij}\right]$	Green–Lagrange strain tensor	(2.24)	29
$\Delta\mathbf{E}^L$	linear strain increment component at step n	(3.33)	61
$\Delta\mathbf{E}^{NL}$	nonlinear strain increment component at step n	(3.34)	61
$E,\ \nu$	Young modulus and Poisson coefficient for elasticity	(2.94)	41
$\mathbf{e} = \left[e_{ij}\right]$	Euler–Almansi strain tensor	(2.18)	28
$\boldsymbol{\epsilon} = \left[\varepsilon_{ij}\right]$	infinitesimal strain tensor	(2.28)	30

ε	infinitesimal uniaxial strain	(4.1)	79
$\mathbf{F}=[F_{ij}]$	deformation gradient tensor	(2.22)	29
$\mathbf{f}^V=[f_i^V]$	body force vector in Lagrangian configuration	(2.51)	34
$\mathbf{f}^v=[f_i^v]$	body force vector in Eulerian configuration	(2.34)	31
$\mathbf{f}^C(P)$	Lagrangian muscle contraction force vector	(6.66)	162
$\mathbf{f}^c(P)$	Eulerian muscle contraction force vector.	(5.41)	136
$f^M(l^M, v^M)$	muscle uniform contraction force.	(5.41)	136
$\mathbf{\Phi}=[\Phi_{ij}]$	material relaxation function.	(2.99)	44
$\mathbf{G}=[G_{ij}]$	reverse deformation gradient tensor.	(2.5)	26
$G(t)$	reduced relaxation function	(4.25)	83
$\mathbf{\Gamma}^v=[\Gamma_i^v]$	acceleration vector of a point	(2.34)	31
$\mathbf{H}=[H_{ij}]$	tensorial shape-function	(3.1)	52
$\mathbf{H}^f=[H_{ij}^f]$	tensorial shape-function for external body force vector	(3.11)	55
$\mathbf{H}^t=[H_{ij}^t]$	tensorial shape-function for external contact force vector.	(3.11)	55
$\mathbf{H}^E$	material stiffness matrix for creep	(2.109)	45
$\mathbf{H}^M$, $\mathbf{D}^M$	tensorial shape functions for the uniform force f^M	(6.70)	164
$\mathbf{h}_n^{P_i}$, $\mathbf{h}_n^{Q_i}$	bulk and shear stress history vectors	(3.61)	68
I_E, II_E, III_E	invariants of $\mathbf{E}$	(2.26)	30
I_C, II_C, III_C	invariants of $\mathbf{C}$	(2.26)	30
I_1, I_2, I_3	invariants of $\mathbf{C}$ in terms of the extension ratios λ_i	(2.27)	30
$J(t)$	reduced creep function	(4.31)	84
$\mathbf{J}$	Jacobian determinant of the deformation.	(2.6)	27
$\mathbf{K}=[K_{ij}]$	stiffness matrix.	(3.19)	57
$\mathbf{K}^E$	material stiffness matrix for linear elasticity	(2.92)	41
$\mathbf{K}^V$	viscous matrix for linear viscoelasticity	(3.56)	66
$\mathbf{K}_n$	tangent stiffness matrix	(3.76)	71
$\tilde{\mathbf{K}}_n$	modal tangent stiffness matrix	(3.91)	75
$\mathbf{K}_n^M$	material tangent stiffness matrix	(3.40)	63
$\mathbf{K}_n^L$	linear tangent stiffness matrix	(3.47)	64
$\mathbf{K}_n^{NL}$	geometric stiffness matrix.	(3.47)	64
$K(\lambda, t)$	material relaxation function.	(4.9)	81
$\mathbf{L}^s=[L_i^s]$	external load at the current boundary surface.	(2.32)	31
$\mathbf{L}=[L_i]$	external load matrix	(3.15)	56
$\mathbf{L}_n$	nodal external force vector at step n	(3.47)	64

L	Lagrange multiplier related to the hydrostatic pressure	(2.85)	40
$\lambda,\ \mu$	Lamé constants for elasticity	(2.91)	41
λ_i	principal extension ratios	(2.16)	28
$\lambda \equiv \lambda_1$	uniaxial extension ratio	(4.2)	79
$\mathbf{M} = [M_{ij}]$	mass matrix	(3.15)	56
$\mathbf{M}_n$	tangent mass matrix	(3.47)	64
$\tilde{\mathbf{M}}_n$	modal tangent mass matrix	(3.91)	75
$\mathbf{N} = [N_i]$	nodal coordinate vector	(3.1)	52
$\mathbf{N}^V$	material damping matrix for creep	(2.109)	45
$\eta_1,\ \eta_2$	Maxwell viscoelasticity constants	(2.109)	45
$\mathbf{P}_i,\ \mathbf{Q}_i$	Lagrangian bulk and shear stress tensors	(2.119)	47
$\mathbf{p}_i,\ \mathbf{q}_i$	Eulerian bulk and shear stress tensors	(2.117)	47
$P_i,\ Q_i$	bulk and shear relaxation moduli	(2.116)	47
$P,\ Q$	bulk an shear relaxation functions	(2.116)	47
p	hydrostatic pressure	(2.79)	39
$\mathbf{\Pi}(\mathbf{U},\dot{\mathbf{U}})$	internal nodal force vector	(3.14)	56
$\boldsymbol{\Psi},\ \boldsymbol{\psi}_k$	modal transfer matrix and eigenvectors at time 0	(3.89)	75
$\mathbf{R} = [R_{ij}]$	rotation component of the deformation	(2.29)	31
$\mathbf{R}_n$	nodal internal force vector	(3.47)	64
ρ	current mass density	(2.34)	31
ρ_0	initial volume mass	(2.40)	32
$\mathbf{S} = [S_{ij}]$	Piola–Kirchhoff second stress tensor	(2.45)	33
$\mathbf{S}'$	distortional component of **S**	(2.83)	39
$\mathbf{S}_n^V$	viscous stress-history influence vector	(3.56)	66
$\mathbf{S}_n^R$	residual stress tensor	(3.40)	63
$s,\ S$	surface in Eulerian, Lagrangian configuration	(2.8)	27
$\boldsymbol{\sigma} = [\sigma_{ij}]$	Cauchy stress tensor	(2.33)	31
$\boldsymbol{\sigma}$	vectorial form of $\boldsymbol{\sigma}$	(2.49)	34
$\boldsymbol{\sigma}'$	distortional component of $\boldsymbol{\sigma}$	(2.79)	39
$\boldsymbol{\sigma}_n^R$	residual stress tensor	(3.49)	65
$\boldsymbol{\sigma}_n^V$	viscous stress-history influence vector	(3.69)	69
$\mathbf{T} = [T_{ij}]$	Lagrange or Piola–Kirchhoff first stress tensor	(2.43)	32
$T \equiv T_{11}$	uniaxial Lagrange stress	(4.2)	79
$T^{(e)}(\lambda)$	elastic response	(4.12)	81
$\mathbf{t}^s = [t_i^s]$	stress vector at the current boundary surface	(2.32)	31
$\mathbf{t}$	time	(2.1)	26

$\theta_\lambda,\ \theta_\mu$	Kelvin–Voigt viscoelasticity constants	(2.103)	44
$\mathbf{U},\ \dot{\mathbf{U}},\ \ddot{\mathbf{U}}$	nodal displacement, velocity, acceleration vector	(3.10)	55
$\mathbf{u}=[u_i]$	displacement vector of point P	(2.19)	29
$\mathbf{u}^f,\ \mathbf{u}^t$	displacement vectors of external body and contact forces	(2.47)	34
$\mathbf{V}_n$	generalized modal displacement vector at step n	(3.89)	75
v, V	volume in Eulerian, Lagrangian configuration	(2.7)	27
W	hyperelasticity strain energy function	(2.63)	37
$W_1,\ W_2$	distortional, dilational components of W	(2.75)	38
$\mathbf{w}_n$	nodal active strength vector at step n	(6.70)	164
$\mathbf{\Omega}$	tensorial shape function for ω	(6.70)	164
$\omega(P)$	fiber active strength distribution function	(5.48)	137
$\mathbf{X}(X_i)$	initial coordinate vector of point P	(2.4)	26
$\mathbf{x}(x_i)$	current coordinate vector of point P	(2.1)	26
$\mathbf{Y}_n$	effective internal force vector at step n	(3.82)	72

1 Soft Tissue Physiology

From a physiological point of view, any solid component of the organism from bones to cells may be considered as a living tissue. Soft tissues may be distinguished from other tissues like bones for their flexibility, their soft mechanical properties. This concerns the connective tissues, the muscles, the organs and the brain (Lee 82). A more accurate distinction may be made in considering their respective functions in the organism. Bones are dedicated to building the rigid skeletal structure of the body, cartilage to lubricating the articulations, skeletal muscles to producing strength and to moving the skeleton through the tendons, and organs and brain play physiological functions to maintain and control the organism. In this report, we are mainly concerned with the mechanical properties of the soft tissues involved in body motion and deformation, i.e., skeletal muscles, tendons, ligaments, and skin. Their respective mechanical behavior may be expected to be related to their specific composition, structure, location, and function in the organism. However, for dynamic modeling, they may be satisfactorily approximated by macroscopic properties, as described in the next chapter.

1.1 Tendons and Ligaments

1.1.1 Composition and Structural Description

Elliott has elaborated a comprehensive report on the microscopic aspects and properties of tendon (Elliott 65). Though their respective mechanical properties and their functions are slightly different, tendons and ligaments show roughly the same material composition (Viidik 87). The prime constituent of these tissues, named *collagen*, represents among 75% dry weight in human tendon (Crisp 72).

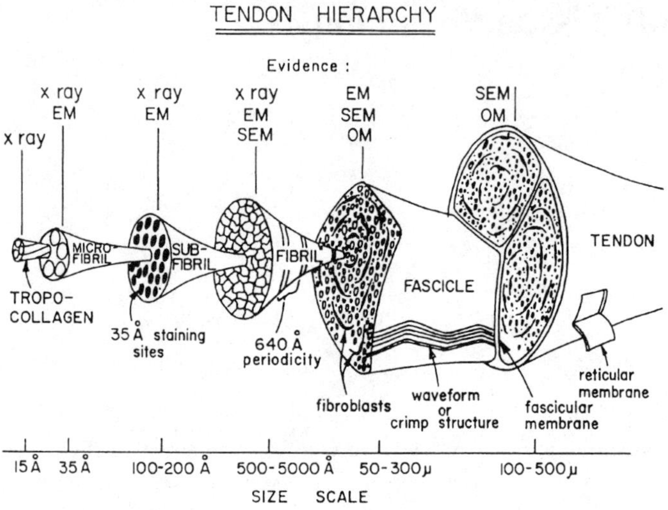

Fig. 1.1. The fibrous structure of tendon
(reprinted from (Kastelic 78) with permission of Gordon and Breach)

The remaining weight is shared between elastin, reticulin, and a hydrophilic gel called *ground substance* (Fung 93b). As described in Fig. 1.1, the tendon fascicles are organized in hierarchical bundles of fibers arranged in a more or less parallel fashion, beginning with *tropocollagen* molecules, which self-assemble into *microfibrils*, which then aggregate to form *subfibrils*, which organize into the structural unit referred to as the *fibril*, the elemental component of the fibers (Woo 93). The diameter of collagen fibrils ranges between 200 Å and 4000 Å depending on the tendon size and function. Their diameter distribution was analyzed for mature rat tail tendon by Parry et al. (Parry 78) and reported to be characterized by two peaks of average diameters 700 Å and 2800 Å (Silver 87).

While the fibers are oriented roughly parallel to the direction of the effort, the tendon fibrils appear in microstructural form in a wave pattern referred to as crimp (Woo 93). However, other soft tissues such as skin, though possessing roughly the same composition as tendon, show large differences in their mechanical response in similar experimental conditions (Crisp 72). This observation has led to the conclusion that the mechanical properties of soft tissues are due rather to their structure than to the relative amount of their constituents (Fung 87). A close look at the fiber network shows that the parallel arrangement of the fibers is more irregular and distributed in more directions for ligaments than for tendons (Fung 93b). More than their chemical composition, the structure of soft tissues is clearly responsible for their distinct classification and function.

1.1.2 Mechanical Properties

Ligaments and tendons fulfill different functions in the organism. The first ones connect bones together, handle the stability of joints and restrict their ranges of motion, while the second ones transmit muscle forces to bones in order to move the skeleton (Viidik 87). One can then conclude that both of them work as links handling and transmitting loads between the skeletal components. This explains the prime interest for characterizing their mechanical properties.

Elasticity. Tendon is an unusual case of living tissue where the state of stress in vivo can be approximated by uniaxial stress-strain relationships (Silver 87). This property is due to the parallel organization of the tendon fibers in the direction of the load to handle. In order to characterize the elasticity of the tissue as a whole, tendon has been submitted to uniaxial tensile tests. The purpose of such experiments is to observe the relationship between the applied load and the resulting extension when static equilibrium is reached. A typical tensile curve of tendon is shown in Fig. 1.2 (Viidik 80).

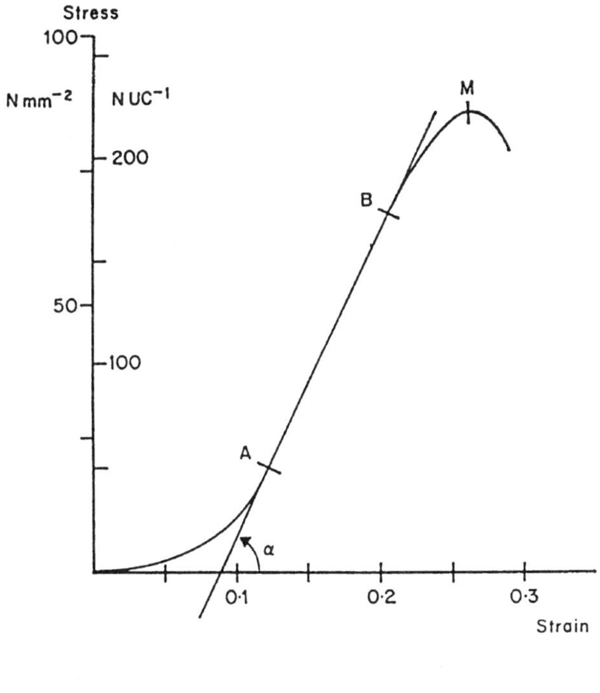

Fig. 1.2. Load-extension curve of tendon
(reprinted from (Viidik 80) with permission of Academic Press)

Kwan described the phenomenon as follows: *Under uniaxial tension, parallel-fibered collagenous tissues exhibit a non-linear stress-strain relationship characterized by an initial low modulus region, an intermediate region of gradually increasing modulus, a region of maximum modulus which remains relatively constant, and a final region of decreasing modulus before complete tissue rupture occurs. The low modulus region is attributed to the removal of the undulations of collagen fibrils that normally exist in a relaxed tissue. As the fibrils start to resist the tensile load, the modulus of the tissue increases. When all the fibrils become taut and loaded, the tissue modulus reaches a maximum value, and thereafter, the tensile stress increases linearly with increasing strain. With further loading, groups of fibrils begin to fail, causing the decrease in modulus until complete tissue rupture occurs* (Kwan 89).

The first parts of the curve are more useful from a functional point of view, since they correspond to the physiological range in which the tissue normally functions (Fung 93b). Tendons are designed to transmit loads with a minimum amount of energy loss and deformation, and are required to stretch less than 10% of their original length (Silver 87). A strain of 4% is regarded as especially significant in tendons and corresponds to the disappearance of fiber waviness (Crisp 72). The tensile strength for whole tendons was reported to be about 150 to 300 MPa for pure collagen (Harkness 68), while the elastic stiffness of a tendon fascicle is about 1000 MPa (Viidik 87).

Viscoelasticity. The above experiment reveals the relationship between stress and strain in the static case. However, when the equilibrium is not reached, a history-dependent component exists in the mechanical behavior of living tissues (Fung 72). When measured in dynamic extension, the stress values appear higher than

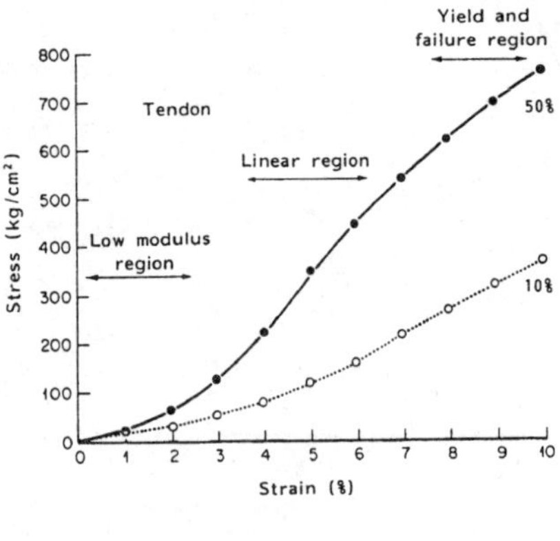

Fig. 1.3. Influence of the strain rate (reprinted from (Birk 91) with permission of Plenum Press)

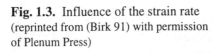

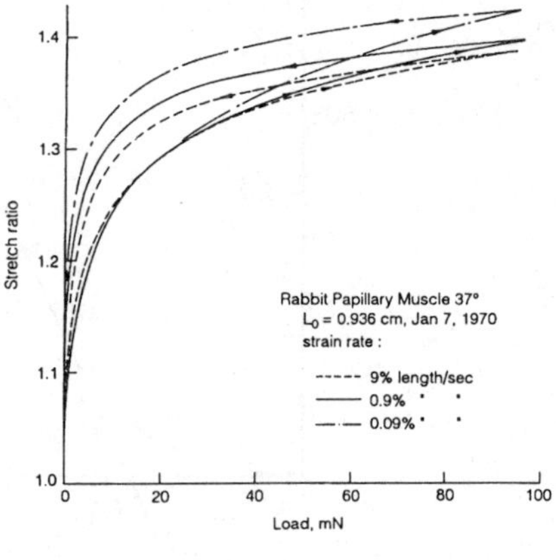

Fig. 1.4. Hysteresis curve (reprinted from (Fung 72) with permission of Prentice Hall)

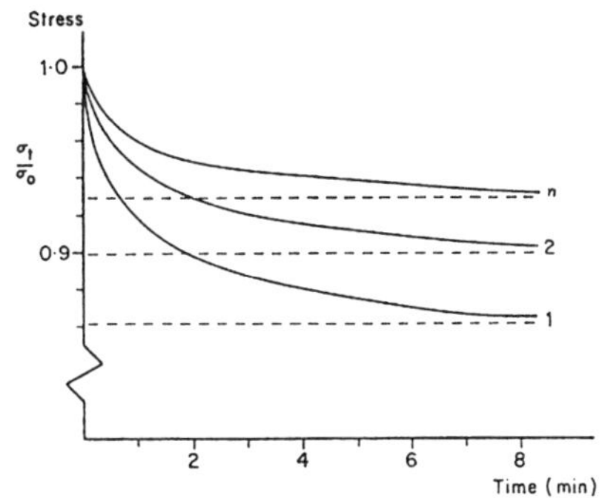

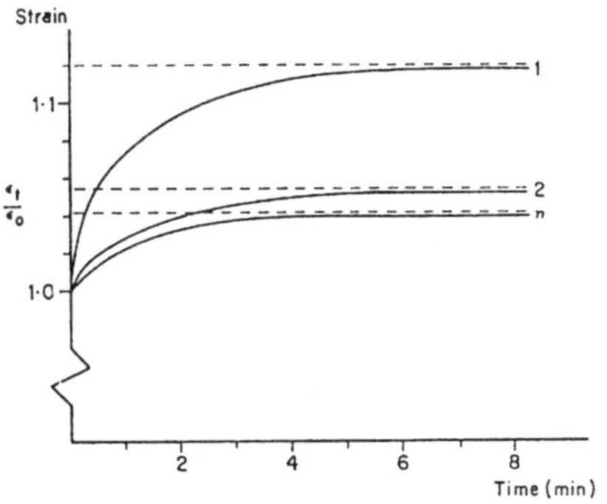

Fig. 1.5. Stress-relaxation curve **Fig. 1.6.** Creep curve
(reprinted from (Viidik 80) with permission of Academic Press)

those at equilibrium, for same strain. The resulting tensile curve appears steeper than the one at equilibrium (Fig. 1.3). When a tissue is suddenly extended and maintained at its new length, the stress gradually decreases slowly against time. This phenomenon is called *stress relaxation* (Fig. 1.5). When the tissue is suddenly submitted to a constant tension, its lengthening velocity decreases against time until equilibrium. This phenomenon is called *creep* (Fig. 1.6). Under cyclic loading, the stress-strain curve shows two distinct paths corresponding to the loading and unloading trajectories. This phenomenon is named *hysteresis* (Fig. 1.4). As global statement, the stress at any instant of time depends not only on the strain at that time, but also on the history of the deformation. These mechanical properties, observed for all living tissues, are common features of a physical phenomenon named *viscoelasticity* (Fung 93b).

1.1.3 Notes on Experimentation

When loading-unloading cycles are applied on the tissue successively up to the same stress level, the stress-strain curve is gradually shifted to the right (Fig. 1.7). After a number of such cycles, the mechanical response of the tissue enters a stationary phase and the results become reproducible from one cycle to the next. This phenomenon is due to the changes occurring in the internal structure of the tissue, until a steady state of cycling is reached. This initial phase of behavior common to all living tissues is usually used as *preconditioning* of the tissues prior to experimentation (Viidik 87). The purpose of testing is to obtain simple, general laws describing the macroscopic behavior of the materials, in order to determine their mechanical properties, and predict their response under defined conditions.

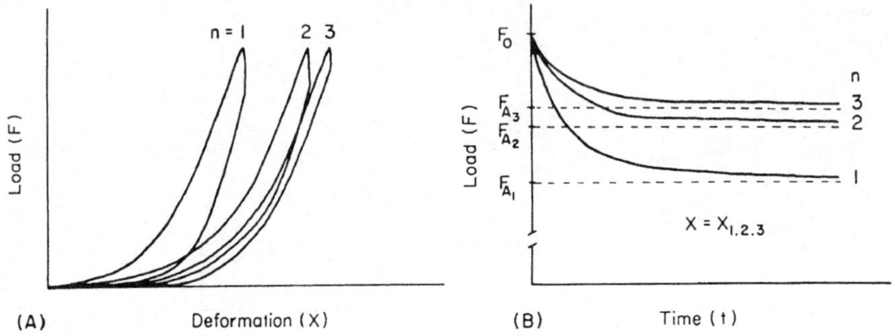

Fig. 1.7. Preconditioning
(reprinted from (Viidik 73) with permission of Academic Press)

The choice of the variables to be measured has thus a prime influence on the modeling process (Fung 87). In usual tensile tests, the forces are referred to the resting section and the extensions to the resting length of the material sample. Graphs are thus plotted for the Lagrangian stress T with respect to the Lagrangian strain ε. Sometimes, this Lagrangian stress is taken for the true stress σ in the resulting constitutive equation. It must be emphasized that this substitution is valid only for strains smaller than 2% of the resting length. However, soft tissues are likely to exceed this limit in their physiological range of functioning, so that, in most cases, this assumption no longer applies. One then talks about *finite strain*, or *large deformation*. In these cases, other variables must be considered in the formulation of the constitutive equations in order to obtain physically valid relations (Chapter 2) (Fung 94).

Whatever the approach adopted, in most cases, the mechanical properties of living materials can hardly be completely assessed. The constants found in one test only apply to the state of deformation which provides these constants, so that no normal response can be uniquely defined for the material. When simple relationships can be written, it is generally only for limited ranges of stresses and strains. Furthermore, the mechanical behavior determined on the basis of experiments cannot be readily correlated with the in vivo conditions of the tissue, where generally the reference state is not completely known (Lee 82). The assumptions grounding the results of the experiments may therefore no longer be valid in the real conditions of functioning (Crisp 72). Finally, experiments are usually limited to a one-dimensional stress field. It is clear that all stresses and strains in the three dimensions are involved, and should be considered for an accurate characterization of living materials. But this would require extensive two- or three-dimensional testing programs, which are difficult to carry out on small, flimsy specimens that must be maintained in living conditions. It is thus difficult to assert the absolute significance of the models developed (Fung 72).

1.2 Skeletal Muscles

1.2.1 Composition and Structural Description

Muscles may be distinguished from the other soft tissues by their specific contractile properties. They function as flexible shortening actuators acting on the different components of the organism. Three types of muscle may be distinguished depending on the functions they fulfill: the skeletal muscles, the smooth muscles, and the heart (Fung 93c). While the skeletal muscles have their ends attached onto separate bones and are responsible of moving of the skeleton, smooth muscles act on organs and other soft tissues like skin in order to fulfill physiological tasks. The heart muscle may be viewed as an organ itself, functioning as a pump irrigating the whole body with blood. All three types of muscle have roughly the same structural and mechanical properties, with differences in size, orientation, strength, and activation properties. Among them all, skeletal muscles are the most powerful ones for handling loads and generating body motion.

When muscle is considered as actuator, not only the contractile fibers are concerned, but also the passive connective tissues tying them to the bones. In a muscle spindle, the fibers are more or less equal in length and uniform in thickness (Kaufman 89). They are arranged in parallel fashion between both tendons of origin and insertion. However, they may appear oriented at an acute angle to the tendons. Such layout is named *pennation* (Zajac 89). Different existing pennate musculo-tendon configurations are shown in Fig. 1.8.

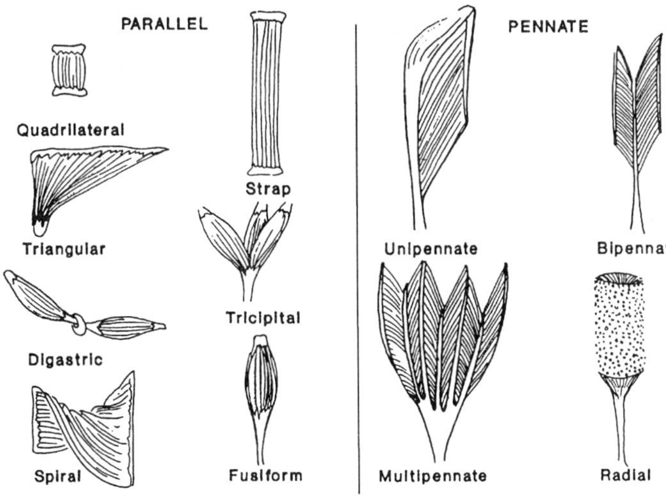

Fig. 1.8. Different musculotendon pennations
(reprinted from (Cutts 93) with permission of Marcel Dekker)

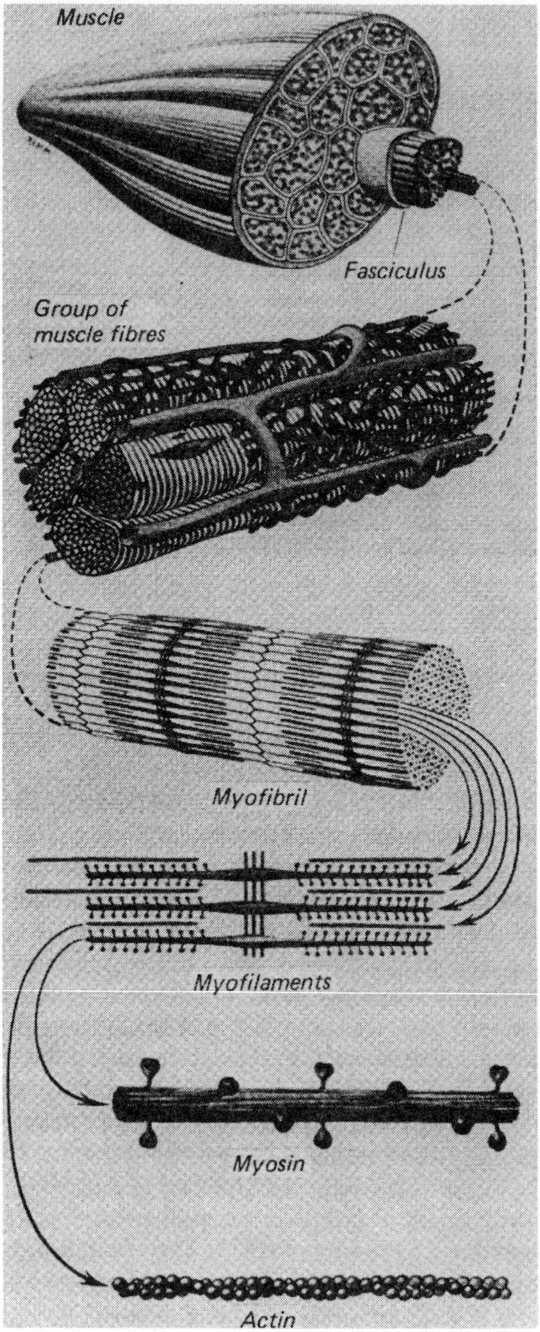

Fig. 1.9. Muscle structure
(reprinted from (Warwick 73) with permission of Churchill Livingstone)

In general, a parallel fibered muscle shortens considerably along the line of action, but exerts only a small force. Such muscles are specifically adapted for movement. Conversely, when the fibers of a pennate muscle contract at an angle to the line of action, very little length change is observed along this line, but considerable force is generated. Such muscle is appropriate for load handling (Cutts 93).

Arranged in a similar fashion as tendon, muscle is composed of fascicles containing bundles of fibers, themselves composed of parallel bundles of *myofibrils*. The *myofibril* is the basic fiber of muscle. It consists of series of contractile units, named *sarcomeres*, composed of arrays of *actin* and *myosin* myofilaments, arranged so as to form the contractile mechanism of muscle. This architecture is shown in Fig. 1.9. The muscle fiber is about $100\ \mu m$ in diameter, the myofibril about $1\ \mu m$, and the myofilaments about $100\ Å$ (Winter 90). The constitutive connective tissue in muscle can be divided into three structural entities, the *epimysium*, which surrounds the whole muscle, the *perimysium*, which surrounds the bundles of muscle fibers, and the *endomysium*, which surrounds each muscle fiber individually (Purslow 89). The ensemble of these tissue sheaths is called the *fascia*. It encloses the muscles, groups or separates them into layers, and ultimately connect them at their ends to the tendons (Winter 90). The *sarcomere* is the basic contractile unit of muscle. The interdigitating thick and thin myofilaments which compose the sarcomere constitute the basis for the sliding filament theory of muscle contraction. The thick filaments are mainly composed of *myosin* molecules arranged in such a manner that a bulbous region of each myosin molecule projects out from the filament. Present understanding is that these bulbous regions extending from the thick *myosin* filaments combine momentarily with the thin *actin* filaments, creating a shearing force between them, which tends to shorten the sarcomere. This connection is called a *cross-bridge* (McMahon 87). The sarcomere length, which is related to the state of activation and the force developed in the muscle, can vary from $1.5\ \mu m$ at full shortening to $2.5\ \mu m$ at resting length, and up to about $4.0\ \mu m$ at full lengthening. The resting length appears here as the optimal length for force generation and motion performance (Winter 90).

1.2.2 Mechanical Properties

It has been shown that the mechanical properties of the connective tissues depend strongly on the orientation of the collagen fiber network. A common feature of such networks is that they are very compliant at small strains but become much stiffer at high strains (Purslow 89). Similar passive load-extension behavior has also been observed for skeletal muscle. Like other tissues, muscle is *non-linear, anisotropic, and viscoelastic* with quite ordinary properties (Fung 93c).

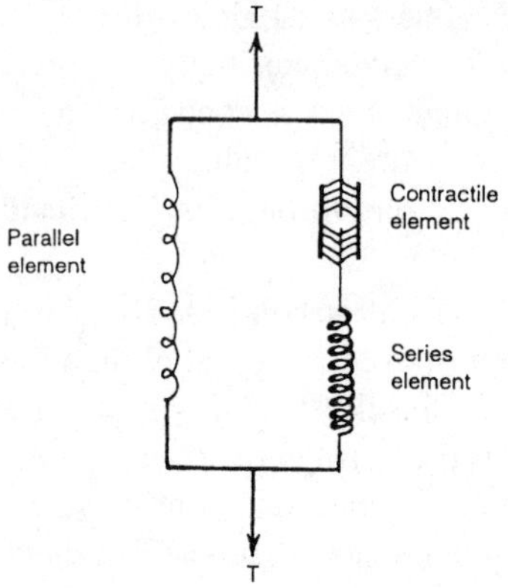

Fig. 1.10. Hill's three-element muscle model
(reprinted from (Fung 93c) with permission of Springer-Verlag)

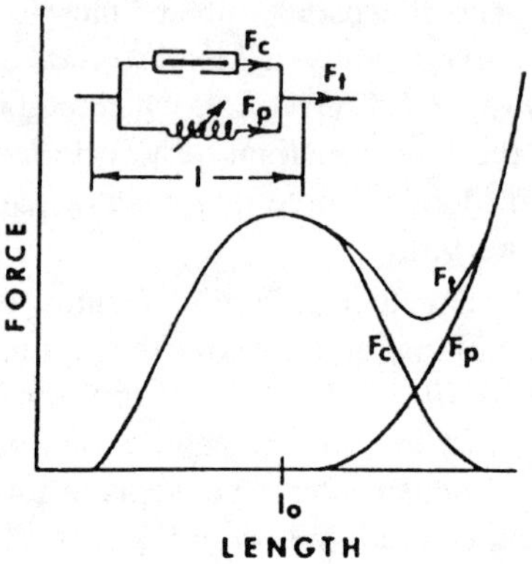

Fig. 1.11. Muscle active and passive force-length relationships
(reprinted from (Winter 90) with permission of John Wiley & Sons)

Long ago, Hill represented the muscle actuator as composed of three elements: two elements arranged in series – one elastic element to account for the muscle elasticity in isometric conditions, and one contractile element, which is freely extendible at rest, but capable of shortening when activated – in parallel with one other elastic element to account for the elasticity of the muscle at rest (Hill 38). While it has been recognized that the parallel element stands for the action of the intramuscular connective tissues surrounding the fibers, the series elastic element has mainly been attributed to the intrinsic elasticity of the cross-bridges (Fig. 1.10) (Fung 93c).

Force-Length Relationship. During isometric contractions (constant muscle length), the series elastic element lengthens slightly as the contractile element shortens. As the muscle lengthens, the parallel element is no longer loose and tension begins to grow up in the non-linear fashion common to all soft tissues. The force-length characteristic of the muscle is a combination of the force-length characteristics of both active and passive components (Winter 90). To draw the force-length characteristic, the passive and fully contracted muscle forces are recorded for a set of static muscle lengths. Then for each length value, the passive force is subtracted from the fully contracted muscle force in order to get the true contraction force developed in the contractile component. Typical isometric active and passive force-length curves are drawn in Fig. 1.11. The shape of the active force-length curve may be explained by the changes occurring in the myofibril structure during contraction. At resting length, there are a maximum number of cross-bridges between filaments, and therefore a maximum tension is possible.

As the muscle lengthens, the filaments are pulled apart, the number of cross bridges reduces, and tension decreases. As the muscle shortens over its resting length, there is an overlapping of the cross-bridges and interferences take place (Winter 90). When the muscle is less than fully activated, the active force-length curve is assumed to be a scaled version of the fully activated one, while the passive relationship, of course, is not affected (Zajac 89). Figure 1.13 shows isometric force-length curves for different activations.

Force-Velocity Relationship. The isometric force-length characteristics, though providing useful information on the contractile properties, do not account for the dynamic properties of muscles. The tension in a muscle decreases as it shortens under load. The reasons for this appear to be the loss in tension as the cross-bridges in the contractile element break and then reform in a shortened condition, as well as the passive damping due to fluid viscosity in both the contractile element and the connective tissues (Winter 90). This phenomenon may be observed in measuring the velocity of shortening of a fully activated muscle submitted to constant loads. A force-velocity graph as shown in Fig. 1.12 may then be drawn to describe the mechanical power output that active muscle delivers (Zajac 89).

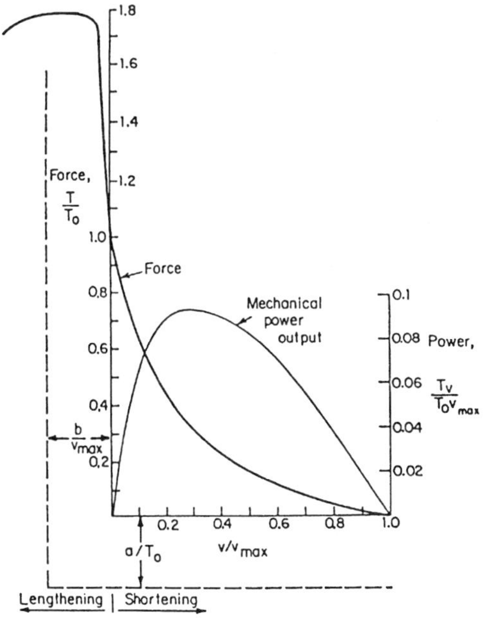

Fig. 1.12. Muscle force-velocity relationships
(reprinted from (Winter 90) with permission of John Wiley & Sons)

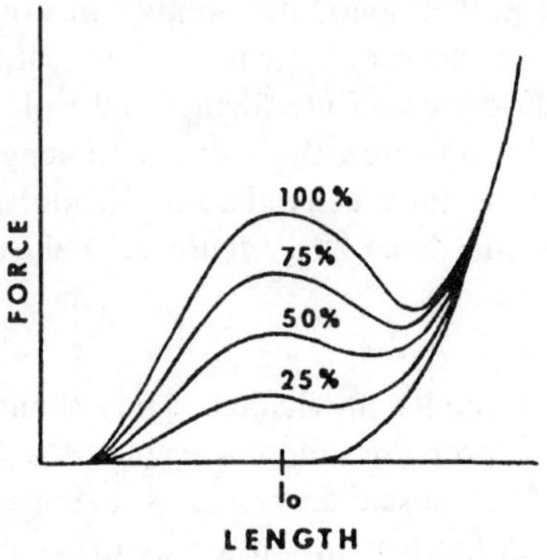

Fig. 1.13. Force-length relationships for different activations

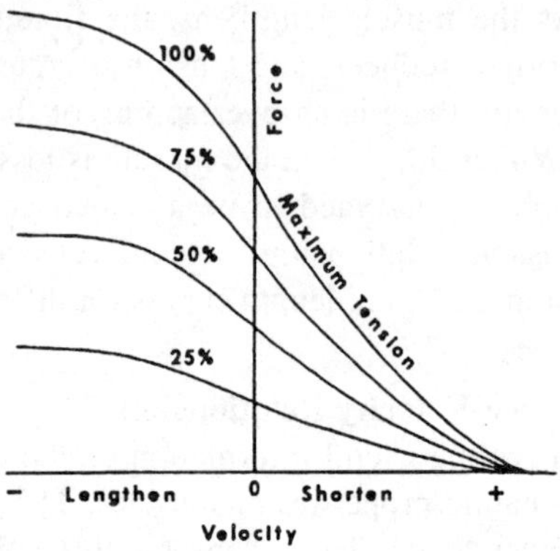

Fig. 1.14. Force-velocity relationships for different activations

(reprinted from (Winter 90) with permission of John Wiley & Sons)

These interpretations are justified by the fact that the curve rate is affected by the activation rate. If only passive effects were involved, the curve would be insensitive to activation changes. If only cross-bridge effects were involved, the force velocity curve would just be scaled by the activation (Winter 90). As shown in Fig. 1.14, the force-velocity curve appears as a combination of both phenomena. As a result of the force-length and force-velocity relationships, the complete mechanical properties of skeletal muscle may be described by a set of three-dimensional surfaces representing the muscle output force as a function of its length, velocity, and activation level (Fig. 1.15).

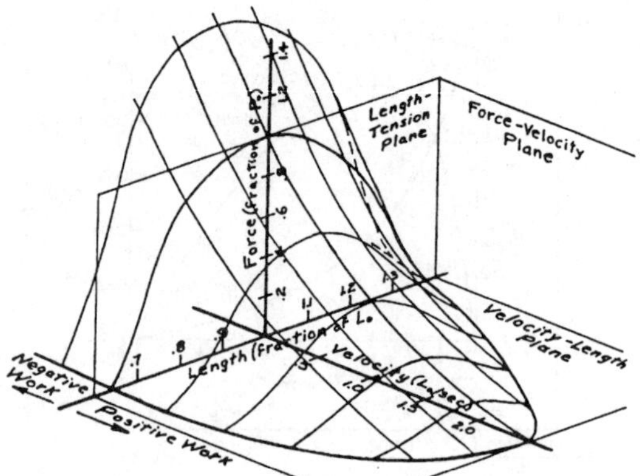

Fig. 1.15. Muscle force-length-velocity graph
(reprinted from (Winter 90) with permission of John Wiley & Sons)

1.2.3 Activation Dynamics

The contraction process is controlled neurologically. The smallest sub-unit that can be controlled is called a *motor unit* because it is innervated separately by a motor axon. Neurologically, the motor unit consists of a synaptic junction in the ventral root of the spinal cord, a motor axon, and a motor end plate in the muscle fibers (Winter 90). Each motor neuron may innervate many muscle fibers. Not all muscle fibers are excited at the same time. The total contraction force of a muscle depends on how many fibers are stimulated. In general, smaller muscles that react rapidly and whose control is exact have small motor units and a large number of nerve fibers, whereas larger muscles which don't require a fine degree of control may have as many as a hundred muscle fibers for each motor unit (Fung 93c).

The basic electrical stimulus of a motor unit is a short duration action potential, which can be considered as an impulse. The mechanical response to this impulse is a much longer duration twitch of tension, which follows quite closely the impulse response of a critically damped second-order system (Fig. 1.16). Muscle activation appears here as an intermediate variable integrating the sequence of neural discharges and describing the level of contraction of the muscle. An increase of tension can therefore be accomplished either by an increase in the stimulation rate for that motor unit or by the recruitment of an additional motor unit (Fig. 1.17). The recruitment of motor units is accomplished according to the size principle, which states that the smaller units are recruited before the larger ones. Conversely, tension is reduced by releasing the motor units from the larger to the smaller ones (Winter 90). The usual technique to represent the motoneuron drive is to record the electromyographic signal of the muscle (EMG measurements). As the neural excitation takes the form of a train of impulses, the usual EMG signal appears as composed of spikes in series. Raw EMG are usually applied for the identification of the muscles acting during motion performance (Fig. 1.18). A logical extension of this is to attempt to extrapolate the force generated by the muscle from the size of the raw EMG.

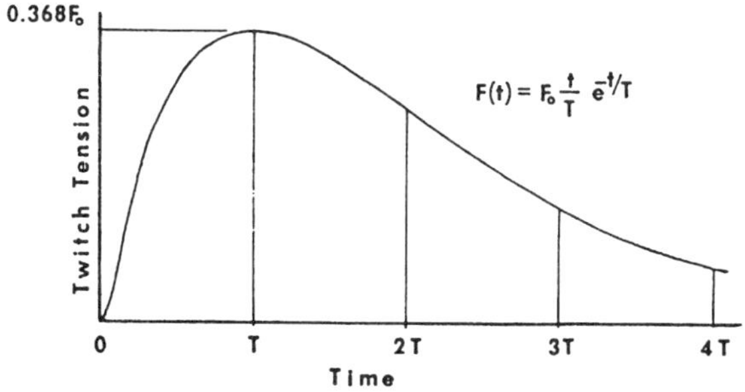

Fig. 1.16. The muscle twitch
(reprinted from (Winter 90) with permission of John Wiley & Sons)

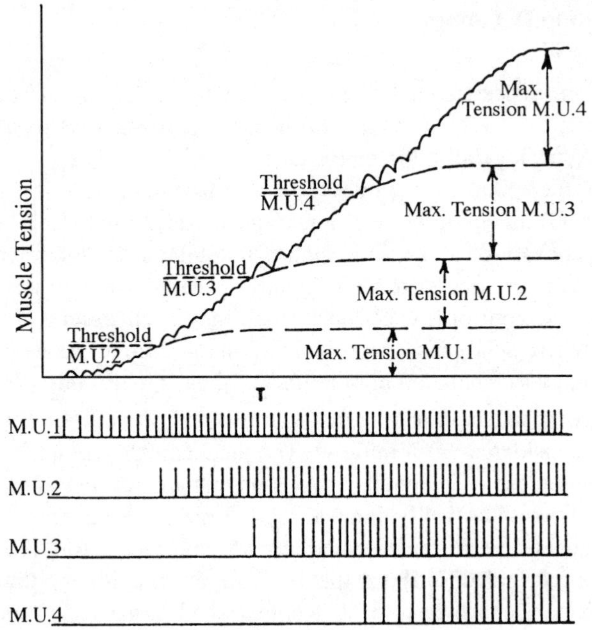

Fig. 1.17. Recruitment of motor units (M.U.)
(reprinted from (Winter 90) with permission of John Wiley & Sons)

The general figure of a EMG-force relationship, in which the force quoted could be realistically attributed to a single muscle, shows a quasi-linear relationship followed by an exponential curve (Fig. 1.19). However, EMG data must be undertaken only with extreme caution because of technical limitations as well as misinterpretations of the results (Cutts 93).

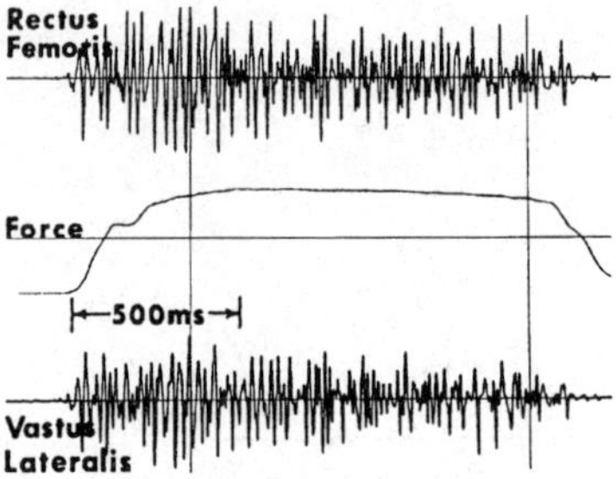

Fig. 1.18. Integrated raw EMG curve (reprinted from (Winter 90) with permission of John Wiley & Sons)

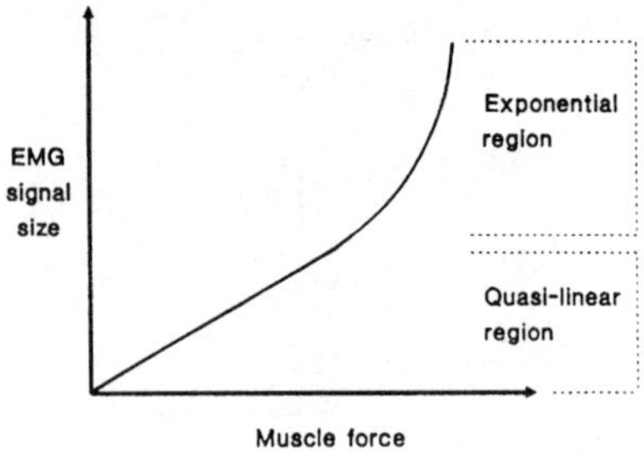

Fig. 1.19. EMG to force relationship (reprinted from (Cutts 93) with permission of Marcel Dekker)

1.3 Skin

1.3.1 Composition and Structural Description

Skin provides the outermost covering of the human body. It forms a barrier between the body and external environment. Mechanically, the most important functions of skin are to support internal organs and protect them from injury while allowing considerable mobility at the same time. In constitutive description, human skin is a *non-homogeneous, anisotropic, non-linear viscoelastic multi-component* material. Skin accounts for about 16 percent of the body weight. Its surface area is 1.5 to 2.0 m^2 in adults, and it varies in thickness from 0.2 (eyelid) to 6.0 mm (sole of foot). It lies on the subcutaneous fatty tissue, which in turn overlies on the densely fibrous fascia. Skin consists of two layers, the *epidermis* and the *dermis.* The epidermis is a relatively thin layer of stratified epithelium. Its thickness varies from 0.07 to 0.12 mm over most of the body but it may reach 0.8 mm on the palms and 1.4 mm on the soles (Lanir 87). The dermis mainly contains randomly oriented collagen and elastic fibers which are embedded in a very viscous matrix, called the *ground substance*. Its thickness cannot be measured exactly as it passes over into the subcutaneous layer without a sharp boundary. In average, it is approximately 1 to 2 mm, but at some regions, it is less than 0.6 mm or reaches 3.0 mm (Bloom 75). A typical cross-section of human skin is shown in Fig. 1.20 (Danielson 73).

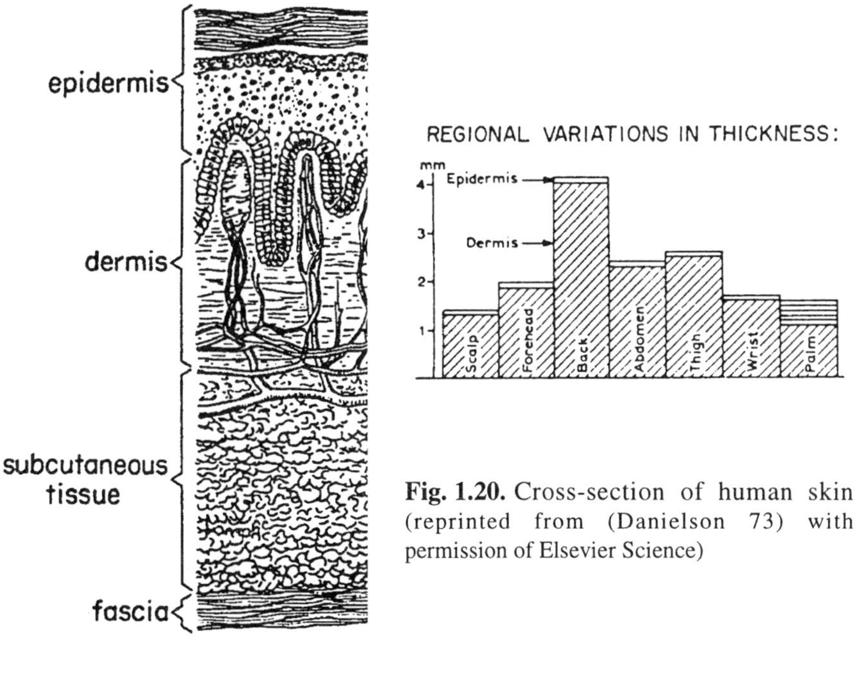

Fig. 1.20. Cross-section of human skin (reprinted from (Danielson 73) with permission of Elsevier Science)

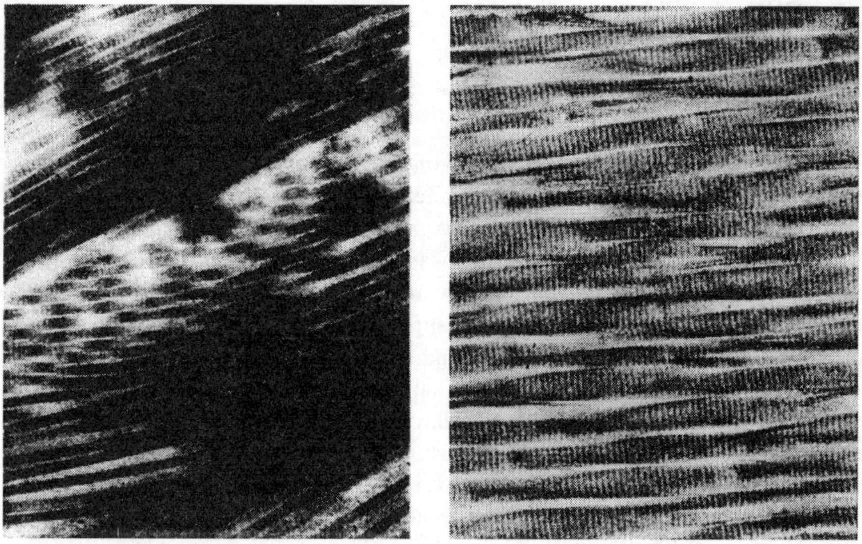

a. Meshwork of collagen fibrils in skin **b.** Parallel collagen fibrils in a tendon

Fig. 1.21. Collagen fiber structure comparison between skin and tendon (reprinted from (Viidik 73) with permission of Academic Press)

Collagen. Collagen fibers are the major constituents in skin and form a three dimensional, disordered network of wavy coiled fibers, with some preferential orientation (Fig. 1.21a) but not as much as in tendon (Fig. 1.21b). In skin, most of the collagen fibers seem to lie in one plane. They are strong (tensile strength of 1.5 to 350 MPa) and stiff (Young modulus in the linear region is approximately 1 GPa) (Lanir 87).

Elastin. Elastin fibers appear fine. They are first stretched when the tissue is strained. It is considerably less stiff than collagen but can be reversibly stretched to more than 100 percent. Elastin plays an important role in the skin response at low strain levels when the collagen fibers are still crimped. The elastin fibers have also some preferential orientation (Lanir 87).

Ground Substance. The ground substance is a gelatinous matrix responsible for the viscoelastic behavior of skin. The fluid-like ground substance exudes from the interfiber space while the fibers become reoriented and densely packed upon the stretch (Lanir 87). Dense connective tissues contain a very small amount of ground substance while loose connective tissues contain a lot (Fung 93b).

1.3.2 Mechanical Properties

In the skin, the fibers are organized in several biaxial layers of thin membranes with some interconnections between them. The important mechanical properties of skin are its extensibility, its resistance to friction, and its response to lateral compressive loading. The first one is mainly related to the dermis, the second one to the epidermis, and the last one to the combined effects of both layers (Lanir 87). These properties vary with species, age, exposure, hydration, obesity, disease, site, and orientation. The changes occurring in the skin after imposed deformation is twofold: on one hand the fibers rotate and stretch, on the other hand, under pressure, the fluid of the matrix is expelled out from between the fibers. This produces the viscoelastic behavior of the skin. The other properties of the skin are described hereafter.

Non-homogeneous, Non-linear. The collagen and elastin fibers can be considered linearly elastic respectively, but the stress-strain relation of the skin is non-linear due to the non-uniformity of its structure (Fig. 1.22). It is usually convenient to discuss the skin deformation in three phases: at very low strains, the response of collagen fibers can be ignored, the elastin fibers are mainly responsible for the stretching of the skin, the stress-strain relation is approximately linear (I); at average strains, a gradually straightening of the originally undulated collagen fibers occurs, which makes the skin tissue become stiffer (II); at high strains, the collagen fibers are all straight, the stress-strain relation becomes linear again (III).

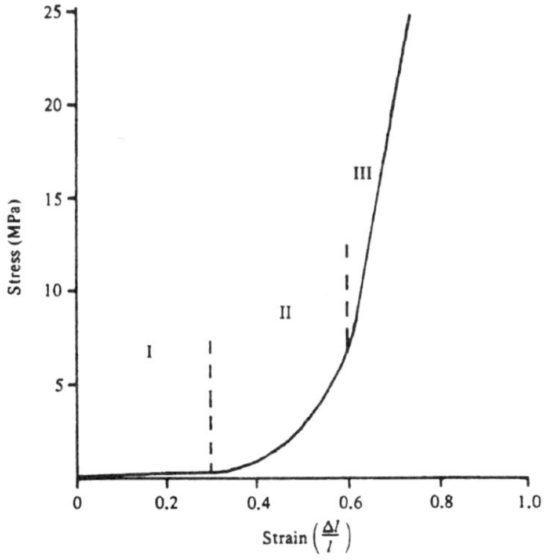

Fig. 1.22. Typical stress-strain curve for skin
(reprinted from (Elden 77) with permission of John Wiley & Sons)

Table 1.1. High modulus strain region (Markenscoff 79)

Strain	Age
0.5–0.65	18 days
0.3–0.50	43 years
0.15–0.20	74 years

Beyond the third phases, the limit strength is reached and the skin fibers begin to break (Veronda 70, Elden 77). As the phases I and III are approximately linear, the skin behavior may be considered as biphasic. Table 1.1 gives for different ages the critical strain for entry into the high modulus region III of the tensile curve of skin (Markenscoff 79). It is observed that this critical strain decreases with age, which can be interpreted as a gradual stretching of the collagen fibers with age.

Anisotropy. Skin is an anisotropic material naturally under tension. The natural lines of skin include the *pre-stress* lines (Langer's line), the *contour* lines (at junctions of skin planes such as where the nose joins the cheek) and the *wrinkle* lines. Langer's cleavage lines are related to the visible crease and wrinkles lines of skin (Fig. 1.23). As they correspond to the directions of preferred disposition of

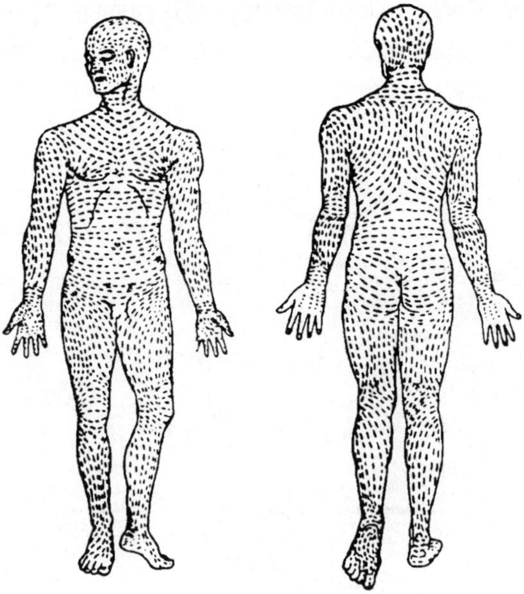

Fig. 1.23. Langer's lines
(reprinted from (Cox 42) with permission of Blackwell Science)

collagen fibers, the skin's mechanical anisotropy follows the same directional pattern. The elastin and collagen fibers along Langer's lines are more stretched than those across the lines, so that the extensibility of skin is lower in the direction of these lines and its stiffness is higher in this direction. But at higher strains the slope in the linear region of the stress-strain curve is similar in all directions. Therefore, though it is not a homogeneous material, in many cases skin can be simplified to be statistically homogeneous.

Viscoelasticity. Skin is a viscoelastic tissue. Its stress-strain relationships are rate-dependent and exhibit considerable *hysteresis*: in a cyclic process, the skin stress-strain relationship in loading is usually somewhat different from that in unloading. Skin exhibits *stress relaxation* under constant strain: when it is suddenly strained and then maintained for a period, the stress induced in the body decrease with time. Skin exhibits *creep* under constant stress: if the skin is suddenly stressed and then the stress is maintained constant, it continues afterwards to deform. The viscoelastic response of the skin is not linear: relaxation and creep depend on the corresponding levels of stress or strain (Fung 93a).

Incompressibility. The volume compressibility of skin has also been investigated. Skin contains a high proportion of chemically bound water molecules as little free fluid inside. The volume compressibility is defined as $\Delta V / V \Delta P$, where ΔV is the volume change, under the pressure change ΔP. The compressibility is $0.38\ m^2GN^{-1}$ for human skin while $0.49\ m^2GN^{-1}$ for water (North 78). Skin is extremely difficult to compress, the change in volume being exceedingly small for large changes in pressure. Therefore, in most cases, it can be considered incompressible. This property is valuable for simplifying the formulation of the constitutive relationship.

Plasticity. If the deformations are small enough, skin can obviously be reversibly stretched. However, when it is stretched frequently or to a relatively large strain for a long period, permanent deformation occur which appears as permanent wrinkles or sutured wounds. Up to now, little is known about this phenomenon.

1.3.3 Connection with Subcutaneous Tissues

Skin is a passive soft tissue, i.e., it deforms following the deformation of the underlying soft tissues. Their mechanical properties have significant influence on the appearance and behavior of skin. Wrinkles, contour lines, and other aging effects involve not only the non-linear elasticity of skin, but also the non-homogeneity of the subcutaneous tissues. Figure 1.24 shows a simplified drawing of the layered structure skin of skin. The first two layers, the epidermis and the

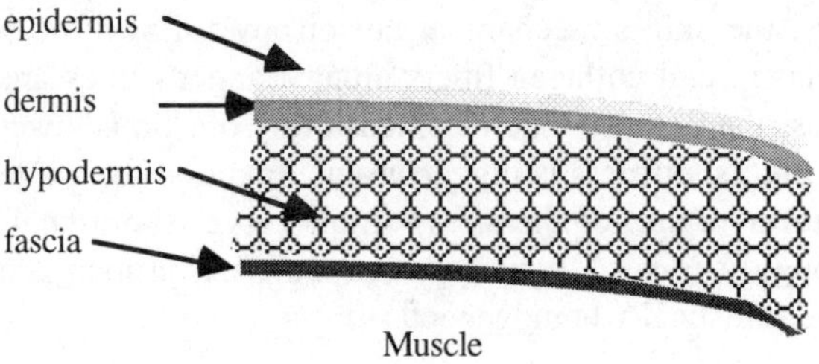

Fig. 1.24. Layered skin structure

dermis, are closely connected together. Meanwhile, the dermis is connected by collagen fibers to a subcutaneous fatty tissue, called the *hypodermis*. The hypodermis appears as a honeycomb structure connected to the fascia which surrounds the muscle bundles. This structure serves as fat container for the body. The binding forces between the skin and the subcutaneous tissues are significant and varies in different sites on the body. It was observed that a slide loading of 8×10^{-3} N is necessary to displace $1.0\ \text{cm}^2$ of human forearm skin by 1.0 mm (Lanir 1987).

Little research has been done regarding the behavior of the subcutaneous tissues. The hypodermis is sometimes considered as a third layer of skin. Its constitution varies largely according to its location or to the individual. It is the hypodermis that provides the skin's loose flexible connection with the other internal soft tissues, whereas the upper layers are more resistant to protect from injuries. Nevertheless, in mechanical experiments, because of its low resistance to tension, the hypodermis is often removed from the upper layers. Since its mechanical behavior is quite different from the one of the upper layers, it seems more appropriate to consider separately the mechanical properties of the upper layers and those of the subcutaneous tissues.

Conclusion

Soft tissues may be regarded as composite materials, clearly non-isotropic because of their fibrous oriented structure. They are non-homogeneous because of their aggregate composition (Crisp 72). Some of them may be regarded as incompressible continua while others are treated as compressible (Lee 82). The mechanical properties previously described lead to identify them as non-linear viscoelastic materials, likely to undergo large deformations when subjected to external constraints (Crisp 72). Before investigating the biomechanical literature searching for constitutive models, we find it appropriate to analyze the meaning of these mechanical properties from the point of view of theoretical mechanics. This is the purpose of the following chapter.

References

This chapter is essentially based on (Crisp 72), (Fung 93abc), (Viidik 87), (Winter 90), (Elden 77), (Lanir 87), (CHARM D4).

Birk 91	D.E. Birk, F.H. Silver, R.L. Trelstad (1991), Matrix assembly, in *Cell Biology of Extracellular Matrix,* 2nd edn. by E.D. Hay. New York: Plenum Press
Bloom 75	W. Bloom, D.W. Fawcett (1975), Skin, in *A Textbook of Histology*. 10th ed. Philadelphia: W.B. Saunders Co.
CHARM D4	W. Maurel, Y. Wu (1994), Survey of mechanical models for tissue deformation and muscle contraction with identification of parametric data, LIG-EPFL/MIRALab-UG, ESPRIT 9036 Project CHARM, Deliverable D4
Cox 42	H.T. Cox (1942), The cleavage lines of skin, *Br. J. Surg.*, 29, 234–240
Crisp 72	J.D.C. Crisp (1972), Properties of tendon and skin, in *Biomechanics: Its Foundations and Objectives*, Y.C. Fung. New York: Prentice-Hall
Cutts 93	A. Cutts (1993), Muscle physiology and electromyography, in *Mechanics of Human Joints: Physiology, Pathophysiology, and Treatment*, ed. by V. Wright, E.L. Radin. New York: Marcel Dekker
Danielson 73	D.A. Danielson (1973), Human skin as an elastic membrane, *J. Biomechanics*, 6, 539–546
Elden 77	H.R. Elden (1977), *Biophysical Properties of Skin.* New York: Wiley-Interscience
Elliott 65	D.H. Elliott (1965), Structure and function of mammalian tendon, *Biol. Rev.* , 40, 392–421
Fung 72	Y.C. Fung (1972), Stress-strain history relations of soft tissues in simple elongation, in *Biomechanics: Its Foundations and Objectives*, Y.C. Fung, N. Perrone, M. Anliker. Englewood Cliffs, NJ: Prentice-Hall
Fung 87	Y.C. Fung (1987), Mechanics of soft tissues, in *Handbook of Bioengineering*, co-ed. by R. Skalak, S. Chien. New York: McGraw-Hill
Fung 93a	Y.C. Fung (1993), The meaning of the constitutive equation, in *Biomechanics: Mechanical Properties of Living Tissues.* Berlin: Springer-Verlag
Fung 93b	Y.C. Fung (1993), Bioviscoelastic solids, in *Biomechanics: Mechanical Properties of Living Tissues.* Berlin: Springer-Verlag

Fung 93c Y.C. Fung (1993), Skeletal muscle, in *Biomechanics: Mechanical Properties of Living Tissues.* Berlin: Springer-Verlag

Fung 94 Y.C. Fung (1994), *A First Course in Continuum Mechanics: for Physical and Biological Engineers and Scientists*, 3rd edn., Englewood Cliffs, NJ: Prentice-Hall

Warwick 73 *Gray's Anatomy*, 35th British edn. by R. Warwick, P. L. Williams (1973). Philadelphia: W.B. Saunders

Harkness 68 R.D. Harkness (1968), Mechanical properties of collagenous tissues, in *Treatise on Collagen, Vol. 2: Biology of Collagen*, ed. by B.S. Gould. New York: Academic Press

Hill 38 A.V. Hill (1938), The heat of shortening and the dynamic constants of muscle, *Proc. R. Soc.*, B. 126, 136–195

Kastelic 78 J. Kastelic, A. Galeski, E. Baer (1978), The multicomposite ultrastructure of tendon, *Connective Tissue Research,* 6, 11–23

Kaufman 89 K.R. Kaufman, K.N. An, E.Y-S. Chao (1989), Incorporation of the muscle architecture into the muscle length-tension relationship, *J. Biomechanics*, 22, 943–948

Kwan 89 M.K. Kwan, S.L.-Y. Woo (1989), A structural model to describe the non-linear stress-strain behavior for parallel-fibered collagenous tissues, *J. Biomech. Engng.*, 111, 361–363

Lanir 87 Y. Lanir (1987), Skin mechanics, in *Handbook of Bioengineering*, ed. by R. Skalak, S. Chien. New York: McGraw-Hill

Lee 82 G.C. Lee, N.T. Tseng (1982), Finite element analysis in soft tissue mechanics, in *Finite Elements in Biomechanics*, ed. by R.H. Gallagher, B.R. Simon, P.C. Johnson, J.F. Gross. Chichester, UK: John Wiley & Sons

Markenscoff 79 X. Markenscoff, I.V. Yannas (1979), On the stress-strain relation for skin, *J. Biomechanics*, 12, 127–129

McMahon 87 T.A. McMahon (1987), Muscle Mechanics, in *Handbook of Bioengineering*, ed. by R. Skalak, S. Chien. New York: McGraw-Hill

North 78 J.F. North, F. Gibson (1978), Volume compressibility of human abdominal skin, *J. Biomechanics*, 11, 203–207

Parry 78 D.A.D. Parry, A.S. Craig, C.R.G. Barnes (1978), Tendon and ligament from the horse: an ultrastructural study of collagen fibrils and elastic fibres as a function of age, *Proc. Roy. Soc. Lond.,* B. 203, 293–303

Purslow 89 P.P. Purslow (1989), Strain induced reorientation of an intramuscular connective tissue network: implications for passive muscle elasticity, *J. Biomechanics*, 22, 21–31

Silver 87 F.H. Silver (1987), *Biological materials: structure, mechanical properties and modeling of soft tissues,* New York: New York University Press

Veronda 70 D.R. Veronda, R.A. Westmann (1970), Mechanical characterization of skin finite deformations, *J. Biomechanics*, 3, 111–124

Viidik 73 A. Viidik (1973), Functional properties of collagenous tissues, *International Review of Connective Tissue Research*, 6, 127–216, ed. by D.A. Hall, D.S. Jackson. New York: Academic Press

Viidik 80 A. Viidik, J. Vuust (1980), *Biology of collagen: proceedings of a symposium*, Aarhus, July/August 1978. London: Academic Press

Viidik 87 A. Viidik (1987), Properties of tendons and ligaments, in *Handbook of Bioengineering*, ed. by R. Skalak, S. Chien. New York: McGraw-Hill

Winter 90 D.A. Winter (1990), Muscle mechanics, in *Biomechanics and Motor Control of the Human Movements,* 2nd edn. New York: Wiley

Woo 93 S.L.-Y. Woo, G.A. Johnson, B.A. Smith (1993), Mathematical modeling of ligaments and tendons, *J. Biomech. Engng.*, 115, 468–473

Zajac 89 F.E. Zajac (1989), Muscle and tendon: properties, models, scaling and application to biomechanics and motor control, in *CRC Critical Reviews in Biomedical Engineering,* 17, 359–411, ed. by J.R. Bourne. Boca Raton: CRC Press

2 Theoretical Mechanics

According to the previous chapter, soft tissues may be regarded as composite, non-isotropic, non-homogeneous, more or less incompressible, finitely deforming, damaging, non-linear viscoelastic materials. From the biomechanical point of view, soft tissue modeling consists of formulating a constitutive equation relating the tension in the material to the deformation it undergoes. Such a relationship may then be used to predict the stress in the tissue for a given deformation, or to simulate its deformation under external loads. The choice of the variables to consider is thus of prime importance at the modeling stage. For the purpose of a 3D application allowing the simulation of any mechanical action on the tissues, it is necessary to apprehend the validity of the models over the limits of the experiments. Here appears the need to review the theoretical mechanics for a better understanding of the available relations. The purpose of this chapter is therefore to remind the theoretical concepts of stress, strain, large deformations, elasticity, and viscoelasticity, with particular attention to their physical meaning.

2.1 Continuum Mechanics

2.1.1 Homogeneous Deformation

Ω is a finite continuous medium in motion between an initial configuration R_0 defined at time $t_0 = 0$, taken as reference configuration and named *Lagrangian* configuration, and any following configuration R_t defined at time t, taken as current configuration and named *Eulerian* configuration. $\mathbf{X}$ and $\mathbf{x}$ are the continuous coordinate vectors of any point P belonging to Ω, defined respectively with respect to the Lagrangian and Eulerian configurations (Fig. 2.1) (Mal 91).

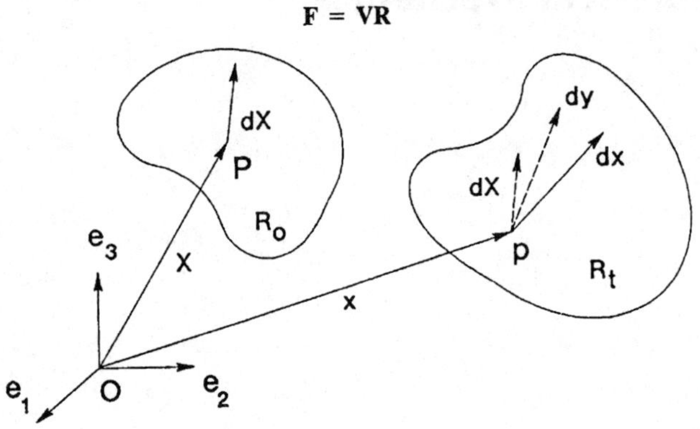

Fig. 2.1. Continuum homogeneous deformation
(reprinted from (Mal 91) with permission of Prentice-Hall)

Deformation Gradient Tensors. The motion of the point P of Ω is described by:

$$\mathbf{x} = \mathbf{x}(\mathbf{X}, t) \qquad \text{or} \qquad x_i = x_i(\mathbf{X}, t) \tag{2.1}$$

Differentiating (2.1) with respect to the Lagrangian coordinates leads to:

$$d\mathbf{x} = \mathbf{F}d\mathbf{X} \qquad \text{or} \qquad dx_i = \sum_j \frac{\partial x_i(\mathbf{X}, t)}{\partial X_j} dX_j \tag{2.2}$$

$$\mathbf{F} = \left[\frac{\partial x_i}{\partial X_j} \right] \tag{2.3}$$

where $\mathbf{F}$ is the *deformation gradient* tensor, which converts any elementary segment $d\mathbf{X}$ of Ω defined with respect to the reference configuration, into a segment $d\mathbf{x}$ defined with respect to the current configuration (Mal 91).

Similarly the *reverse deformation gradient* tensor may be defined as:

$$\mathbf{X} = \mathbf{X}(\mathbf{x}, t) \qquad \text{or} \qquad X_i = X_i(\mathbf{x}, t) \tag{2.4}$$

$$\mathbf{G} = \left[\frac{\partial X_i}{\partial x_j} \right] \qquad \text{i.e.:} \qquad \mathbf{G} = \mathbf{F}^{-1} \tag{2.5}$$

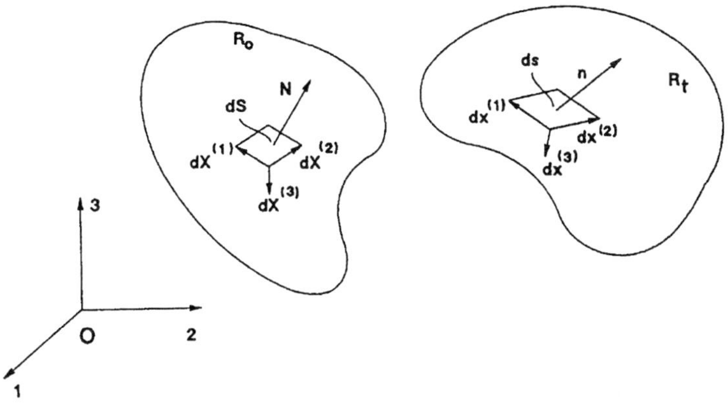

Fig. 2.2. The Eulerian and Lagrangian configurations
(reprinted from (Mal 91) with permission of Prentice-Hall)

Volume Change. The elementary volume dv in the current configuration is:

$$dv = JdV \qquad \text{with} \qquad J = \det(\mathbf{F}) \tag{2.6}$$

$$dv = dx_1 dx_2 dx_3 \qquad \text{and} \qquad dV = dX_1 dX_2 dX_3 \tag{2.7}$$

where dV is the elementary volume in the reference configuration, and **J** the *Jacobian* of the deformation gradient tensor **F** (Fig. 2.2) (Mal 91).

Surface Change. The elementary surface ds in the current configuration is:

$$\mathbf{n}ds = J\mathbf{G}^T\mathbf{N}dS \tag{2.8}$$

where **n** and **N** are the unit vectors, normal to the surface elements ds and dS of the current and reference configurations respectively (Fig. 2.2) (Mal 91).

2.1.2 Strain Analysis

The squared lengths of both segments d**x** and d**X** may be developed as follows:

$$d\mathbf{x}^2 = d\mathbf{x}^T d\mathbf{x} = d\mathbf{X}^T\mathbf{F}^T\mathbf{F}d\mathbf{X} \tag{2.9}$$

$$d\mathbf{X}^2 = d\mathbf{X}^T d\mathbf{X} = d\mathbf{x}^T\mathbf{G}^T\mathbf{G}d\mathbf{x} \tag{2.10}$$

Strain Tensors. The tensorial product upcoming in (2.9) as:

$$\mathbf{C} = \mathbf{F}^{T}\mathbf{F} \tag{2.11}$$

is defined as the *Cauchy–Green right dilation* tensor.

Similarly, the *Cauchy–Green left dilation* tensor may be introduced as:

$$\mathbf{B} = \mathbf{F}\mathbf{F}^{T} \tag{2.12}$$

This allows to define the *Cauchy strain* tensor as:

$$\mathbf{c} = \mathbf{G}^{T}\mathbf{G} = \left(\mathbf{F}^{-1}\right)^{T}\left(\mathbf{F}^{-1}\right) = \left(\mathbf{F}\mathbf{F}^{T}\right)^{-1} = \mathbf{B}^{-1} \tag{2.13}$$

Using (2.9) and (2.11), the squared length variation may be expressed in Lagrangian coordinates as:

$$\mathbf{dx}^{2} - \mathbf{dX}^{2} = \mathbf{dX}^{T}(\mathbf{C}-\mathbf{I})\mathbf{dX} = \mathbf{dX}^{T}2\mathbf{E}\mathbf{dX} \qquad \text{with} \tag{2.14}$$

$$\mathbf{E} = \frac{1}{2}(\mathbf{C}-\mathbf{I}) \tag{2.15}$$

where $\mathbf{E}$ is the *Green–Lagrange* (or Lagrangian) *strain* tensor.

The extension ratios λ_i usually considered in the experiments are related to the principal strain components C_{ii} and E_{ii} by:

$$\lambda_i = F_{ii} = \frac{\partial x_i}{\partial X_i} \qquad C_{ii} = \lambda_i^2 \qquad E_{ii} = \frac{1}{2}\left(\lambda_i^2 - 1\right) \qquad i = 1,2,3 \tag{2.16}$$

Using (2.10) and (2.13), the squared length variation may be expressed in Eulerian coordinates as:

$$\mathbf{dx}^{2} - \mathbf{dX}^{2} = \mathbf{dx}^{T}(\mathbf{I}-\mathbf{c})\mathbf{dx} = \mathbf{dx}^{T}2\mathbf{e}\mathbf{dx} \qquad \text{with} \tag{2.17}$$

$$\mathbf{e} = \frac{1}{2}(\mathbf{I}-\mathbf{c}) \tag{2.18}$$

where $\mathbf{e}$ is the *Euler–Almansi* (or Eulerian) *strain* tensor.

$\mathbf{C}$, $\mathbf{B}$, $\mathbf{c}$, $\mathbf{E}$ and $\mathbf{e}$ are all *second-order symmetric* strain tensors. $\mathbf{C}$ and $\mathbf{E}$ allow the calculation of the length variation of any segment $\mathrm{d}\mathbf{X}$ known in the reference configuration, whereas $\mathbf{c}$ and $\mathbf{e}$ allow the calculation of the length variation of any segment $\mathrm{d}\mathbf{x}$ known in the current configuration (Mal 91).

Strain-Displacement Relationship. In practice, the trajectory function (2.1) of any point of the continuum is unknown configurations (Mal 91). The notion of displacement is thus preferably used:

$$\mathbf{u} = \mathbf{x} - \mathbf{X} \tag{2.19}$$

is the displacement vector of a point P of Ω between both.

Using (2.3) and (2.5), the elementary displacement variation may be defined as:

$$\mathrm{d}\mathbf{u} = \mathrm{d}\mathbf{x} - \mathrm{d}\mathbf{X} = (\mathbf{F} - \mathbf{I})\mathrm{d}\mathbf{X} = (\mathbf{I} - \mathbf{G})\mathrm{d}\mathbf{x} \tag{2.20}$$

which may be expressed in terms of the *displacement gradient* tensor $\nabla\mathbf{u}$:

$$\mathrm{d}\mathbf{u} = \nabla\mathbf{u}\mathrm{d}\mathbf{X} \qquad \text{i.e.:} \qquad \mathrm{du}_i = \sum_j \frac{\partial \mathrm{u}_i}{\partial \mathrm{X}_j}\mathrm{dX}_j \tag{2.21}$$

From (2.11), (2.15), (2.20) and (2.21), comes:

$$\mathbf{F} = \mathbf{I} + \nabla\mathbf{u} \tag{2.22}$$

$$\mathbf{C} = \mathbf{F}^{\mathrm{T}}\mathbf{F} = \mathbf{I} + \nabla\mathbf{u} + \nabla\mathbf{u}^{\mathrm{T}} + \nabla\mathbf{u}^{\mathrm{T}}\nabla\mathbf{u} \tag{2.23}$$

$$\mathbf{E} = \frac{1}{2}\left(\nabla\mathbf{u} + \nabla\mathbf{u}^{\mathrm{T}} + \nabla\mathbf{u}^{\mathrm{T}}\nabla\mathbf{u}\right) \qquad \text{i.e.:} \tag{2.24}$$

$$\begin{cases} \mathrm{E}_{11} = \dfrac{\partial \mathrm{u}_1}{\partial \mathrm{X}_1} + \dfrac{1}{2}\displaystyle\sum_k \left(\dfrac{\partial \mathrm{u}_k}{\partial \mathrm{X}_1}\right)^2 & \mathrm{E}_{12} = \dfrac{1}{2}\left[\dfrac{\partial \mathrm{u}_1}{\partial \mathrm{X}_2} + \dfrac{\partial \mathrm{u}_2}{\partial \mathrm{X}_1} + \displaystyle\sum_k \left(\dfrac{\partial \mathrm{u}_k}{\partial \mathrm{X}_1}\dfrac{\partial \mathrm{u}_k}{\partial \mathrm{X}_2}\right)\right] \\ \mathrm{E}_{22} = \dfrac{\partial \mathrm{u}_2}{\partial \mathrm{X}_2} + \dfrac{1}{2}\displaystyle\sum_k \left(\dfrac{\partial \mathrm{u}_k}{\partial \mathrm{X}_2}\right)^2 & \mathrm{E}_{13} = \dfrac{1}{2}\left[\dfrac{\partial \mathrm{u}_1}{\partial \mathrm{X}_3} + \dfrac{\partial \mathrm{u}_3}{\partial \mathrm{X}_1} + \displaystyle\sum_k \left(\dfrac{\partial \mathrm{u}_k}{\partial \mathrm{X}_1}\dfrac{\partial \mathrm{u}_k}{\partial \mathrm{X}_3}\right)\right] \\ \mathrm{E}_{33} = \dfrac{\partial \mathrm{u}_3}{\partial \mathrm{X}_3} + \dfrac{1}{2}\displaystyle\sum_k \left(\dfrac{\partial \mathrm{u}_k}{\partial \mathrm{X}_3}\right)^2 & \mathrm{E}_{23} = \dfrac{1}{2}\left[\dfrac{\partial \mathrm{u}_2}{\partial \mathrm{X}_3} + \dfrac{\partial \mathrm{u}_3}{\partial \mathrm{X}_2} + \displaystyle\sum_k \left(\dfrac{\partial \mathrm{u}_k}{\partial \mathrm{X}_2}\dfrac{\partial \mathrm{u}_k}{\partial \mathrm{X}_3}\right)\right] \end{cases}$$

Knowing $\mathbf{u}$ allows thus the calculation of $\mathbf{E}$ whichever the coordinate system is.

Invariants. k_1, k_2, k_3 are the three eigenvalues, and I_K, II_K, III_K the three invariants of any second-order tensor $\mathbf{K}$. The invariants of $\mathbf{K}$ may be defined by:

$$\begin{cases} I_K = \mathrm{Tr}(\mathbf{K}) = k_1 + k_2 + k_3 \\ II_K = \frac{1}{2}\left[\mathrm{Tr}(\mathbf{K})^2 - \mathrm{Tr}(\mathbf{K}^2)\right] = k_1k_2 + k_1k_3 + k_2k_3 \\ III_K = \det(\mathbf{K}) = k_1k_2k_3 \end{cases} \tag{2.25}$$

Using (2.15) and (2.25), the invariants I_C, II_C, III_C of $\mathbf{C}$ may be expressed in terms of the invariants I_E, II_E, III_E of $\mathbf{E}$ (Mal 91) as follows:

$$\begin{cases} I_C = 3 + 2I_E \\ II_C = 3 + 4I_E + 4II_E \\ III_C = 1 + 2I_E + 4II_E + 8III_E \end{cases} \tag{2.26}$$

For purpose of experimentation, I_C, II_C and III_C are more useful in terms of the extension ratios λ_1, λ_2, λ_3 than in terms of the principal strains E_{ii}:

$$\begin{cases} I_C = \lambda_1^2 + \lambda_2^2 + \lambda_3^2 & \equiv I_1 \\ II_C = \lambda_1^2\,\lambda_2^2 + \lambda_1^2\,\lambda_3^2 + \lambda_2^2\,\lambda_3^2 & \equiv I_2 \\ III_C = \lambda_1^2\,\lambda_2^2\,\lambda_3^2 & \equiv I_3 \end{cases} \tag{2.27}$$

Linearization. When strains are small, the Lagrangian and Eulerian strain tensors are close enough to allow the common linear approximation:

$$\mathbf{E} \approx \mathbf{e} \approx \boldsymbol{\epsilon} = \frac{1}{2}\left(\nabla\mathbf{u} + \nabla\mathbf{u}^T\right) \qquad \text{i.e.:} \tag{2.28}$$

$$\begin{cases} E_{11} \approx e_{11} \approx \varepsilon_{11} \approx \dfrac{\partial u_1}{\partial X_1} \approx \lambda_1 - 1 & E_{12} \approx e_{12} \approx \varepsilon_{12} \approx \dfrac{1}{2}\left(\dfrac{\partial u_1}{\partial X_2} + \dfrac{\partial u_2}{\partial X_1}\right) \\ E_{22} \approx e_{22} \approx \varepsilon_{22} \approx \dfrac{\partial u_2}{\partial X_2} \approx \lambda_2 - 1 & E_{13} \approx e_{13} \approx \varepsilon_{13} \approx \dfrac{1}{2}\left(\dfrac{\partial u_1}{\partial X_3} + \dfrac{\partial u_3}{\partial X_1}\right) \\ E_{33} \approx e_{33} \approx \varepsilon_{33} \approx \dfrac{\partial u_3}{\partial X_3} \approx \lambda_3 - 1 & E_{23} \approx e_{23} \approx \varepsilon_{23} \approx \dfrac{1}{2}\left(\dfrac{\partial u_2}{\partial X_3} + \dfrac{\partial u_3}{\partial X_2}\right) \end{cases}$$

Polar Decomposition. Small strain does not imply small displacement. This may happen when the strain remains negligible while the rotation is large (Pipkin 86). This involves decomposing the deformation into *rotation* and *distortion*.

Using the polar decomposition property, the deformation gradient tensor $\mathbf{F}$ may be written as composed of an orthogonal *rotation* tensor $\mathbf{R}$, and a *right* ($\mathbf{U}$) or *left* ($\mathbf{V}$) *stretch* tensor corresponding to the pure strain (Ogden 84):

$$\mathbf{F} = \mathbf{R}\mathbf{U} \qquad \text{and} \qquad \mathbf{F} = \mathbf{V}\mathbf{R} \tag{2.29}$$

where $\mathbf{U}$ and $\mathbf{V}$ are respectively related to the right ($\mathbf{C}$) and left ($\mathbf{B}$) Cauchy dilation tensors by:

$$\mathbf{C} = \mathbf{U}^2 \qquad \text{and} \qquad \mathbf{B} = \mathbf{V}^2 \tag{2.30}$$

In case of rigid motion, for example rotation and/or translation, the displacement may be important while the deformation is null. In this case, it comes:

$$\mathbf{U} = \mathbf{V} = \mathbf{C} = \mathbf{B} = \mathbf{c} = \mathbf{I} \qquad \mathbf{F} = \mathbf{R} \qquad \mathbf{E} = \mathbf{e} = \boldsymbol{\epsilon} = [0] \tag{2.31}$$

2.1.3 Stress Analysis

The continuous medium Ω is assumed evolving passively under external loads, from the initial unstressed and undeformed reference state (Trompette 92).

Cauchy Stress. For continuity reasons, the current internal stress vector $\mathbf{t}^s$ along the surface normal $\mathbf{n}$ of Ω is related to the external load $\mathbf{L}^s$ applied on ds by:

$$\mathbf{t}^s(\mathbf{x}, \mathbf{n}) = \lim_{ds \to 0} \frac{d\mathbf{L}^s}{ds} \tag{2.32}$$

According to *Cauchy's theorem*, the stress vector $\mathbf{t}^s$ is related to an internal *second order* stress tensor $\boldsymbol{\sigma}$, corresponding to the physically *true stress* in the material and named *Cauchy's stress* tensor:

$$\mathbf{t}^s(\mathbf{x}, \mathbf{n}) = \boldsymbol{\sigma}^T \mathbf{n} \tag{2.33}$$

Equations of Motion. The dynamics of the medium follow as:

Force balance:
$$\int_v \rho \boldsymbol{\Gamma}^v \, dv = \int_v \mathbf{f}^v \, dv + \int_s \mathbf{t}^s \, ds \tag{2.34}$$

Moment balance: $$\int_v \rho \boldsymbol{\Gamma}^v \wedge \mathbf{x}\, dv = \int_v \mathbf{f}^v \wedge \mathbf{x}\, dv + \int_s \mathbf{t}^s \wedge \mathbf{x}\, ds \tag{2.35}$$

where $\boldsymbol{\Gamma}^v$ is the acceleration of point P, and $\mathbf{f}^v$ and $\mathbf{t}^s$ the external forces. Using the divergence theorem on (2.33) so that:

$$\int_s \mathbf{t}^s\, ds = \int_s \boldsymbol{\sigma}^T \mathbf{n}\, ds = \int_v \mathrm{div}\left(\boldsymbol{\sigma}^T\right) dv \tag{2.36}$$

the equation of motion of Ω may be derived from (2.34) as:

$$\rho \boldsymbol{\Gamma}^v = \mathbf{f}^v + \mathrm{div}\left(\boldsymbol{\sigma}^T\right) \tag{2.37}$$

while (2.35) leads to the symmetry property of the Cauchy stress tensor:

$$\boldsymbol{\sigma}^T = \boldsymbol{\sigma} \tag{2.38}$$

Using (2.38), (2.37) may then as well be written in the form:

$$\rho \boldsymbol{\Gamma}^v = \mathbf{f}^v + \mathrm{div}(\boldsymbol{\sigma}) \qquad \text{or} \qquad \rho \frac{\partial^2 u_i}{\partial t^2} = f_i + \sum_j \frac{\partial \sigma_{ij}}{\partial x_j} \tag{2.39}$$

Kirchhoff Stresses. In practice, the current Eulerian configuration is unknown, so that the equation of motion must be expressed using variables defined with respect to the known initial Lagrangian configuration (Mal 91). This is achieved using (2.6), (2.8), (2.33) and applying the *mass conservation* principle, so that:

$$J = \det(\mathbf{F}) = \frac{dv}{dV} = \frac{\rho_0}{\rho} \tag{2.40}$$

$$\int_V \left(\rho_0 \boldsymbol{\Gamma}^V - \mathbf{f}^V J\right) dV = \int_S J \boldsymbol{\sigma}^T \mathbf{G}^T \mathbf{N}\, dS \qquad \text{i.e.:} \tag{2.41}$$

$$\int_V \left(\rho_0 \boldsymbol{\Gamma}^V - \mathbf{f}^V J\right) dV = \int_S \mathbf{T}^T \mathbf{N}\, dS \qquad \text{with} \tag{2.42}$$

$$\mathbf{T} = J \mathbf{G} \boldsymbol{\sigma} \tag{2.43}$$

where the stress tensor $\mathbf{T}$, appearing in (2.42), obtained from the Cauchy stress tensor $\boldsymbol{\sigma}$ (2.43), corresponds to the internal stress tensor that would be obtained by applying the current external load onto the initial undeformed configuration.

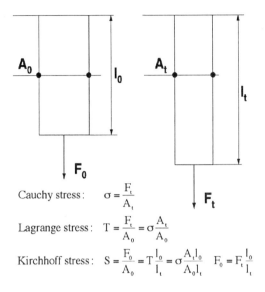

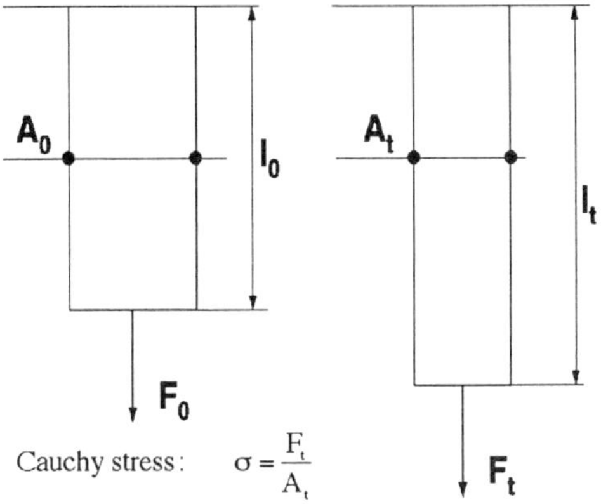

Cauchy stress: $\sigma = \dfrac{F_t}{A_t}$

Lagrange stress: $T = \dfrac{F_t}{A_0} = \sigma \dfrac{A_t}{A_0}$

Kirchhoff stress: $S = \dfrac{F_0}{A_0} = T \dfrac{l_0}{l_t} = \sigma \dfrac{A_t l_0}{A_0 l_t} \quad F_0 = F_t \dfrac{l_0}{l_t}$

Fig. 2.3. Uniaxial definition of the different stresses

$\mathbf{T}$ is named the *Piola–Kirchhoff first stress* tensor, or *Lagrange* stress tensor. It is this stress which is usually considered in the experiments when the load is referred to the initial cross section of the specimen (Fig. 2.3). The use of this tensor is difficult in practice because it is not symmetric (Mal 91). Another stress tensor is rather used, which is symmetric and obtained from the previous one by:

$$\mathbf{S} = \mathbf{T}\mathbf{G}^{\mathrm{T}} = \mathrm{J}\mathbf{G}\boldsymbol{\sigma}\mathbf{G}^{\mathrm{T}} \tag{2.44}$$

This stress tensor is named the *Piola–Kirchhoff second stress* tensor. It corresponds to the stress tensor in the initial configuration, equivalent to the current stress tensor, i.e., it takes into account all the geometrical changes between both configurations (Mal 91). The Lagrangian equation of current motion then becomes:

$$\int_V \left(\rho_0 \boldsymbol{\Gamma}^{\mathrm{V}} - \mathbf{f}^{\mathrm{V}} \mathrm{J}\right) \mathrm{dV} = \int_S \mathbf{F}\mathbf{S}^{\mathrm{T}}\mathbf{N}\, \mathrm{dS} \tag{2.45}$$

2.1.4 Virtual Work Principle

Due to a lack of resolution methods, some problems cannot be solved using the local Eulerian dynamics defined by (2.39). Another mechanical principle must then be used in order to get a suitable equation of motion (Trompette 92).

Eulerian Description. Applying a local virtual displacement $\delta\hat{\mathbf{u}}$ to the volume element, we get from (2.37) the local equation of *virtual work*:

$$\delta\hat{\mathbf{u}}^{T}\rho\boldsymbol{\Gamma}^{v} = \delta\hat{\mathbf{u}}^{T}\mathbf{f}^{v} + \delta\hat{\mathbf{u}}^{T}\text{div}\left(\boldsymbol{\sigma}^{T}\right) \tag{2.46}$$

which, after global integration and transformations, becomes the *variational form*:

$$\int_{v}\delta\hat{\mathbf{u}}^{T}\rho\boldsymbol{\Gamma}^{v}\,dv = \int_{v}\delta\hat{\mathbf{u}}^{f^{T}}\mathbf{f}^{v}\,dv + \int_{s}\delta\hat{\mathbf{u}}^{t^{T}}\mathbf{t}^{s}\mathbf{n}\,ds - \int_{v}\text{Tr}\left(\delta\hat{\mathbf{e}}^{T}\boldsymbol{\sigma}\right)dv \tag{2.47}$$

This formulation is said *weak* or *global* since it approximates the solution of the local problem by the solution of the global problem (Trompette 92). (2.47) is usually written in the form of the *virtual work* principle:

$$\delta\hat{W}^{acc} = \delta\hat{W}^{ext} + \delta\hat{W}^{int} \tag{2.48}$$

with $\delta\hat{W}^{acc} = \int_{v}\delta\hat{\mathbf{u}}^{T}\rho\boldsymbol{\Gamma}^{v}\,dv$ — inertial virtual work

$\delta\hat{W}^{int} = -\int_{v}\text{Tr}\left(\delta\hat{\mathbf{e}}^{T}\boldsymbol{\sigma}\right)dv = -\int_{v}\delta\hat{\boldsymbol{e}}^{T}\boldsymbol{\sigma}\,dv$ — internal virtual work

$\delta\hat{W}^{ext} = \int_{v}\delta\hat{\mathbf{u}}^{f^{T}}\mathbf{f}^{v}\,dv + \int_{s}\delta\hat{\mathbf{u}}^{t^{T}}\mathbf{t}^{s}\mathbf{n}\,ds$ — external virtual work

where $\boldsymbol{\sigma}$, $\boldsymbol{e}$ are the *vectorial* forms of the tensors $\boldsymbol{\sigma}$ and $\mathbf{e}$ defined as:

$$\boldsymbol{\sigma} = \left[\sigma_{11}\quad\sigma_{22}\quad\sigma_{33}\quad\sigma_{i<j}\right]^{T} \quad\text{and}\quad \boldsymbol{e} = \left[e_{11}\quad e_{22}\quad e_{33}\quad 2e_{i<j}\right]^{T} \tag{2.49}$$

The factor 2 in the expression of $\boldsymbol{e}$ has been introduced to account for the duplicate products $\sigma_{ij}e_{ji}$ in $\text{Tr}\left(\delta\hat{\mathbf{e}}^{T}\boldsymbol{\sigma}\right)$, which result from the symmetry of $\boldsymbol{\sigma}$, $\mathbf{e}$.

Lagrangian Description. Using (2.5), (2.15), (2.18), and (2.44), the equivalent state variables in the Eulerian and Lagrangian configurations may be related with:

$$\delta\mathbf{e} = \mathbf{G}^{T}\delta\mathbf{E}\mathbf{G} \quad\text{and}\quad \boldsymbol{\sigma} = J^{-1}\mathbf{F}\mathbf{S}\mathbf{F}^{T} \tag{2.50}$$

Using (2.50), (2.47) may be converted into Lagrangian formulation as:

$$\int_{V}\delta\hat{\mathbf{u}}^{T}\rho_{0}\boldsymbol{\Gamma}^{V}\,dV = \int_{V}\delta\hat{\mathbf{u}}^{f^{T}}\mathbf{f}^{V}\,dV + \int_{S}\delta\hat{\mathbf{u}}^{t^{T}}\mathbf{t}^{S}\mathbf{N}\,dS - \int_{V}\text{Tr}\left(\delta\hat{\mathbf{E}}^{T}\mathbf{S}\right)dV \tag{2.51}$$

where $\mathbf{\Gamma}^{V}$, $\mathbf{f}^{V}$, $\mathbf{t}^{S}$ are the equivalents in the reference configuration of the current acceleration $\mathbf{\Gamma}^{v}$ and external actions $\mathbf{f}^{v}$ and $\mathbf{t}^{s}$ used in the Eulerian description. (2.51) may be written in the usual form of the *virtual work* principle as well:

$$\delta\hat{W}^{acc} = \delta\hat{W}^{ext} + \delta\hat{W}^{int} \tag{2.52}$$

with

$$\delta\hat{W}^{acc} = \int_V \delta\hat{\mathbf{u}}^T \rho_0 \mathbf{\Gamma}^V \, dV \qquad \text{inertial virtual work}$$

$$\delta\hat{W}^{int} = -\int_V \mathrm{Tr}\left(\delta\hat{\mathbf{E}}^T \mathbf{S}\right) dV = -\int_V \delta\hat{\boldsymbol{E}}^T \boldsymbol{S} \, dV \qquad \text{internal virtual work}$$

$$\delta\hat{W}^{ext} = \int_V \delta\hat{\mathbf{u}}^{f^T} \mathbf{f}^V \, dV + \int_S \delta\hat{\mathbf{u}}^{t^T} \mathbf{t}^S \mathbf{N} \, dS \qquad \text{external virtual work}$$

where $\boldsymbol{S}$, $\boldsymbol{E}$ are the *vectorial* forms of the tensors $\mathbf{S}$ and $\mathbf{E}$, defined as:

$$\boldsymbol{S} = \begin{bmatrix} S_{11} & S_{22} & S_{33} & S_{i<j} \end{bmatrix}^T \quad \text{and} \quad \boldsymbol{E} = \begin{bmatrix} E_{11} & E_{22} & E_{33} & 2E_{i<j} \end{bmatrix}^T \tag{2.53}$$

Linear Description. When small deformation may be assumed (geometric linearity), the Eulerian and Lagrangian formulations blend into the same relation:

$$\int_V \delta\hat{\mathbf{u}}^T \rho_0 \mathbf{\Gamma}^V \, dV = \int_V \delta\hat{\mathbf{u}}^{f^T} \mathbf{f}^V \, dV + \int_S \delta\hat{\mathbf{u}}^{t^T} \mathbf{t}^S \mathbf{N} \, dS - \int_V \mathrm{Tr}\left(\delta\hat{\boldsymbol{\epsilon}}^T \boldsymbol{\sigma}\right) dV \tag{2.54}$$

which may be written in the usual form of the *virtual work* principle as:

$$\delta\hat{W}^{acc} = \delta\hat{W}^{ext} + \delta\hat{W}^{int} \tag{2.55}$$

with

$$\delta\hat{W}^{acc} = \int_V \delta\hat{\mathbf{u}}^T \rho_0 \mathbf{\Gamma}^V \, dV \qquad \text{inertial virtual work}$$

$$\delta\hat{W}^{int} = -\int_V \mathrm{Tr}\left(\delta\hat{\boldsymbol{\epsilon}}^T \boldsymbol{\sigma}\right) dV = -\int_V \delta\hat{\boldsymbol{\epsilon}}^T \boldsymbol{\sigma} \, dV \qquad \text{internal virtual work}$$

$$\delta\hat{W}^{ext} = \int_V \delta\hat{\mathbf{u}}^{f^T} \mathbf{f}^V \, dV + \int_S \delta\hat{\mathbf{u}}^{t^T} \mathbf{t}^S \mathbf{N} \, dS \qquad \text{external virtual work}$$

where $\boldsymbol{\sigma}$, $\boldsymbol{\epsilon}$ are the *vectorial* forms of the tensors $\boldsymbol{\sigma}$ and $\boldsymbol{\epsilon}$, defined as:

$$\boldsymbol{\sigma} = \begin{bmatrix} \sigma_{11} & \sigma_{22} & \sigma_{33} & \sigma_{i<j} \end{bmatrix}^T \quad \text{and} \quad \boldsymbol{\epsilon} = \begin{bmatrix} \varepsilon_{11} & \varepsilon_{22} & \varepsilon_{33} & 2\varepsilon_{i<j} \end{bmatrix}^T \tag{2.56}$$

As the current state is unknown, only (2.52), (2.55) are suitable forms for solution.

2.2 Elasticity

2.2.1 Non-linear Elasticity

The previous section considered stress and strain as state variables in omitting their interdependencies. The specific deformation response of a material under stress may be described by its *constitutive relationship* in the form of a function involving the state variables. Usually, a material is said to be *elastic* when the stress is related to the strain. More precisely, the theory of elasticity states that the stress at any time t depends only upon the local deformation at that time, and not upon the history of the deformation (Oden 89). The formulation for elasticity in large deformation may be established as a relation between the deformation gradient tensor **F** and the Piola–Kirchhoff second stress tensor **S** as:

$$\mathbf{S} = \mathbf{S}(\mathbf{F}) \tag{2.57}$$

In order to satisfy the *material indifference* principle, which establishes that the internal state of the material must be invariant under any rigid body deformation, **S** should depend on **F** through the *Cauchy–Green* right dilation tensor **C**:

$$\mathbf{S} = \mathbf{S}(\mathbf{C}) \tag{2.58}$$

Isotropy. In the case of an isotropic material, $\mathbf{S}(\mathbf{C})$ may be developed as follows:

$$\mathbf{S}(\mathbf{C}) = a_0\mathbf{I} + a_1\mathbf{C} + .. + a_n\mathbf{C}^n \tag{2.59}$$

Then, using the *Cayley–Hamilton* theorem, which states that the tensor **C** satisfies its own *characteristic equation*:

$$\mathbf{C}^3 - I_C\mathbf{C}^2 + II_C\mathbf{C} - III_C\mathbf{I} = \mathbf{0} \tag{2.60}$$

the general constitutive equation for elasticity (2.57) becomes:

$$\mathbf{S}(\mathbf{C}) = a_0\mathbf{I} + a_1\mathbf{C} + a_2\mathbf{C}^2 \tag{2.61}$$

where a_0, a_1, a_2 are functions of the invariants I_C, II_C, III_C, or with (2.15):

$$\mathbf{S}(\mathbf{E}) = \alpha_0\mathbf{I} + \alpha_1\mathbf{E} + \alpha_2\mathbf{E}^2 \tag{2.62}$$

These relations include *physical* non-linearities within the coefficients a_0, a_1, a_2 or α_0, α_1, α_2 and within the quadratic tensorial terms in **C** or **E**, as well as *geometrical* non-linearities since **C** or **E** are non-linear tensors with respect to **F**.

2.2.2 Hyperelasticity

An homogeneous, continuous material is said *hyperelastic* if the Piola–Kirchhoff first stress tensor **T** is derived from an internal strain energy function W(**F**) as:

$$\mathbf{T} = \frac{\partial W(\mathbf{F})}{\partial \mathbf{F}} \tag{2.63}$$

Like above, accounting for the *material indifference* property, which introduces **F** through **C**, the constitutive relationship for hyperelasticity becomes:

$$\mathbf{S} = \mathbf{T}\mathbf{G}^{T} = \frac{\partial W(\mathbf{C})}{\partial \mathbf{C}}\frac{\partial \mathbf{C}}{\partial \mathbf{F}}\mathbf{F}^{-T} \qquad \text{i.e.:} \qquad \mathbf{S} = 2\frac{\partial W(\mathbf{C})}{\partial \mathbf{C}} = \frac{\partial W(\mathbf{E})}{\partial \mathbf{E}} \tag{2.64}$$

Isotropy. If the material is isotropic, the strain energy W may be expressed in terms of the three invariants of **C** only:

$$W(\mathbf{C}) = W(I_C, II_C, III_C) \tag{2.65}$$

This allows **S** to be developed in the form:

$$\mathbf{S} = 2\left[W_I \frac{\partial I_C}{\partial \mathbf{C}} + W_{II}\frac{\partial II_C}{\partial \mathbf{C}} + W_{III}\frac{\partial III_C}{\partial \mathbf{C}}\right] \qquad \text{with} \tag{2.66}$$

$$W_I = \frac{\partial W(\mathbf{C})}{\partial I_C} \qquad W_{II} = \frac{\partial W(\mathbf{C})}{\partial II_C} \qquad W_{III} = \frac{\partial W(\mathbf{C})}{\partial III_C} \tag{2.67}$$

Using (2.33), the derivatives of the invariants may be obtained as:

$$\frac{\partial I_C}{\partial \mathbf{C}} = \mathbf{I} \qquad \frac{\partial II_C}{\partial \mathbf{C}} = I_C\mathbf{I} - \mathbf{C} \qquad \frac{\partial III_C}{\partial \mathbf{C}} = II_C\mathbf{I} - I_C\mathbf{C} + \mathbf{C}^2 \tag{2.68}$$

which leads again to the general form of elasticity:

$$\mathbf{S}(\mathbf{C}) = a_0\mathbf{I} + a_1\mathbf{C} + a_2\mathbf{C}^2 \tag{2.69}$$

with coefficients for hyperelasticity of the form:

$$a_0 = 2\left(W_I + I_C W_{II} + II_C W_{III}\right) \quad a_1 = -2\left(W_{II} + I_C W_{III}\right) \qquad a_2 = 2W_{III} \tag{2.70}$$

Using (2.15), this may also be expressed in the form:

$$\mathbf{S}(\mathbf{E}) = \alpha_0\mathbf{I} + \alpha_1\mathbf{E} + \alpha_2\mathbf{E}^2 \tag{2.71}$$

Hyperelasticity is thus a particular case of elasticity (Ciarlet 86). As in Sect. 2.2.1, *geometrical* and *physical* non-linearities remain included within the coefficients, the quadratic tensorial terms in **C** or **E**, and the non-linearities of **C** or **E** versus **F**.

The Strain Energy Function. For purpose of experimentation, the strain energy functions are more appropriate expressed using the strain invariants $I_C = I_1$, $II_C = I_2$, $III_C = I_3$ in terms of the extension ratios λ_1, λ_2, λ_3 (2.27). In an undeformed state, $I_1 = 3$, $I_2 = 3$, and $I_3 = 1$. The general form of the strain energy may then be written as an infinite series in power of $(I_1 - 3)$, $(I_2 - 3)$, and $(I_3 - 1)$ (Ogden 84):

$$W(I_1, I_2, I_3) = \sum_{p,q,r=0}^{\infty} c_{pqr}\,(I_1 - 3)^p\,(I_2 - 3)^q\,(I_3 - 1)^r \qquad \begin{cases} c_{000}, a_{000} = 0 \\ (p,q,r) \text{ integers} \end{cases} \tag{2.72}$$

$$W = \sum_{p,q,r=0}^{\infty} a_{pqr}\left\{\left[\lambda_1^p\left(\lambda_2^q + \lambda_3^q\right) + \lambda_2^p\left(\lambda_3^q + \lambda_1^q\right) + \lambda_3^p\left(\lambda_2^q + \lambda_1^q\right)\right]\left(\lambda_1\lambda_2\lambda_3\right)^r - 6\right\} \tag{2.73}$$

This allows us to derive the principal components of the Lagrangian stress **T** as:

$$T_{ii} = \frac{2}{\lambda_i}\left[\lambda_i^2 \frac{\partial W}{\partial I_1} + I_2 \frac{\partial W}{\partial I_2} - \frac{I_3}{\lambda_i^2}\frac{\partial W}{\partial I_2} + I_3 \frac{\partial W}{\partial I_3}\right] \qquad i = 1,2,3 \ \text{(Allaire 77)} \tag{2.74}$$

The third invariant I_3 being involved in volume properties, W is assumed to be composed of a *distortional* energy W_1 and a *dilational* energy W_2 as:

$$W(I_1, I_2, I_3) = W_1(I_1, I_2) + W_2(I_3) \qquad \text{with} \qquad \begin{cases} c_{000}, a_{000} = 0 \\ (p,q,r) \text{ integers} \end{cases} \tag{2.75}$$

$$W_1(I_1, I_2) = \sum_{p,q=0}^{\infty} c_{pq0} \, (I_1 - 3)^p \, (I_2 - 3)^q \qquad W_2(I_3) = \sum_{r=1}^{\infty} c_{00r} \, (I_3 - 1)^r \tag{2.76}$$

$$W_1 = \sum_{p=1}^{\infty} 2a_{p00} \, (\lambda_1^p + \lambda_2^p + \lambda_3^p - 3) \qquad W_2 = \sum_{r=1}^{\infty} 6 \, a_{00r} \left[(\lambda_1 \lambda_2 \lambda_3)^r - 1 \right] \tag{2.77}$$

2.2.3 Incompressibility

For many solids, the change in volume at large deformations is small compared to the change in shape. Such materials are said to be *incompressible* (Oden 89). This property is expressed using the *mass conservation* principle (2.40):

$$\frac{dv}{dV} = \frac{\rho_0}{\rho} = J = \det(\mathbf{F}) = 1 \tag{2.78}$$

For an isotropic incompressible solid subjected to a *hydrostatic pressure* p, the true stress may be decomposed into elastic and pressure stress components as:

$$\boldsymbol{\sigma} = -p \, \mathbf{I} + \boldsymbol{\sigma}'(\mathbf{C}) \tag{2.79}$$

and the equivalent Kirchhoff stress tensor $\mathbf{S}$ may be written as:

$$\mathbf{S} = J\mathbf{G}\boldsymbol{\sigma}\mathbf{G}^T = -pJ\mathbf{G}\mathbf{I}\mathbf{G}^T + J\mathbf{G}\boldsymbol{\sigma}'(\mathbf{C})\mathbf{G}^T \qquad \text{i.e.:} \qquad \mathbf{S} = -pJ\mathbf{C}^{-1} + \mathbf{S}'(\mathbf{C}) \tag{2.80}$$

Hyperelasticity. As we have from (2.6), (2.11), (2.25), (2.60), and (2.68):

$$III_C = \det(\mathbf{C}) = \det(\mathbf{F}^T\mathbf{F}) = J^2 \tag{2.81}$$

$$\frac{\partial III_C}{\partial \mathbf{C}} = III_C \mathbf{C}^{-1} \tag{2.82}$$

if the material is hyperelastic, the Kirchhoff stress tensor $\mathbf{S}$ becomes:

$$\mathbf{S} = -\frac{p}{J} \frac{\partial III_C}{\partial \mathbf{C}} + \mathbf{S}'(\mathbf{C}) \tag{2.83}$$

Reminding (2.75), the dilational energy function W_2 may thus be written using a *Lagrange multiplier* L and (2.83) becomes:

$$\mathbf{S} = \frac{\partial W_1(I_C, II_C)}{\partial \mathbf{E}} + \frac{\partial W_2(III_C)}{\partial \mathbf{E}} \qquad \text{with} \quad W_2(III_C) = L(III_C - 1) \quad \text{i.e.:} \tag{2.84}$$

$$\mathbf{S} = \frac{\partial W_1(I_C, II_C)}{\partial \mathbf{E}} + L\frac{\partial III_C}{\partial \mathbf{E}} \qquad \text{with} \qquad L = -\frac{p}{2J} \tag{2.85}$$

2.2.4 Linear Elasticity

For a continuous, homogeneous, isotropic, elastic material, the constitutive relationship given by (2.61) still contains combined *geometrical* and *physical* non-linearities.

Geometrical Linearity. This may be assumed when the deformations are small compared to the initial shape of the body. The Lagrangian and Eulerian configurations are then very close, and the common assumptions:

$$\mathbf{E} \approx \mathbf{e} \approx \boldsymbol{\epsilon} \qquad \text{and} \qquad \mathbf{S} \approx \mathbf{T} \approx \boldsymbol{\sigma} \tag{2.86}$$

lead to approximate (2.61) and (2.62) with:

$$\boldsymbol{\sigma} \approx a_0\mathbf{I} + a_1\mathbf{C} + a_2\mathbf{C}^2 \qquad \text{and} \tag{2.87}$$

$$\boldsymbol{\sigma} \approx \alpha_0\mathbf{I} + \alpha_1\boldsymbol{\epsilon} + \alpha_2\boldsymbol{\epsilon}^2 \tag{2.88}$$

which relates the true stress $\boldsymbol{\sigma}$ to the Lagrangian strain $\boldsymbol{\epsilon}$ with a *physically non-linear* relation (non-linear coefficients, quadratic tensorial terms) (Leipholz 74).

Physical Linearity. Physical linearity is obtained using (2.26) and (2.15) in developing the tensor **C** and its invariants I_C, II_C, III_C, up to the first order in terms of the tensor **E** and its invariants I_E, II_E, III_E, as follows:

$$\begin{cases} \mathbf{C} = \mathbf{I} + 2\mathbf{E} \\ \mathbf{C}^2 = \mathbf{I} + 4\mathbf{E} \\ \mathbf{C}^3 = \mathbf{I} + 6\mathbf{E} \end{cases} \qquad \begin{cases} I_C = 3 + 2I_E \\ II_C = 3 + 4I_E \\ III_C = 1 + 2I_E \end{cases} \tag{2.89}$$

As a result, (2.61) takes the form:

$$\mathbf{S} = \mathbf{S}^0 + \lambda I_E\mathbf{I} + 2\mu\mathbf{E} \tag{2.90}$$

Since the initial state of the material is assumed unstressed and undeformed $\mathbf{S} = \mathbf{E} = [\mathbf{0}]$ gives $\mathbf{S}^0 = [0]$ and the *physically linear* relation (2.90) becomes:

$$\mathbf{S} = \lambda \mathrm{Tr}(\mathbf{E})\mathbf{I} + 2\mu\mathbf{E} \tag{2.91}$$

where λ and μ are named the *Lamé constants*. In this case, the equation relating the tensors $\mathbf{S}$ and $\mathbf{E}$ is a linear law, but both Lagrangian stress and strain tensors remain geometrically non-linear due to the large deformation (Leipholz 74).

Combined Linearities. When both geometrical and physical linearities can be assumed, both previous simplifications are valid, and their application results in the usual linear elastic constitutive relationship, the generalized *Hooke* law:

$$\boldsymbol{\sigma} = \mathbf{K}^{\mathrm{E}}\boldsymbol{\epsilon} = \lambda \mathrm{Tr}(\boldsymbol{\epsilon})\,\mathbf{I} + 2\mu\,\boldsymbol{\epsilon} \qquad \text{i.e.:} \tag{2.92}$$

$$\boldsymbol{\sigma} = \boldsymbol{K}^{\mathrm{E}}\boldsymbol{\epsilon} = \lambda \mathrm{Tr}(\boldsymbol{\epsilon})\,\mathbf{I} + 2\mu\,\boldsymbol{\epsilon} \qquad \text{using } \boldsymbol{\sigma} \text{ and } \boldsymbol{\epsilon} \text{ as defined in (2.49):}$$

$$\begin{bmatrix} \sigma_{11} \\ \sigma_{22} \\ \sigma_{33} \\ \sigma_{i<j} \end{bmatrix} = \begin{bmatrix} \lambda+2\mu & \lambda & \lambda & 0 \\ \lambda & \lambda+2\mu & \lambda & 0 \\ \lambda & \lambda & \lambda+2\mu & 0 \\ 0 & 0 & 0 & \mu \end{bmatrix} \begin{bmatrix} \varepsilon_{11} \\ \varepsilon_{22} \\ \varepsilon_{33} \\ 2\varepsilon_{i<j} \end{bmatrix} \tag{2.93}$$

Other forms of this law have been developed, especially to be applied to rheological experiments (Leipholz 74). For example, introducing the *Young modulus* E and *Poisson coefficient* ν, (2.93) may also be written in the forms:

$$\begin{bmatrix} \sigma_{11} \\ \sigma_{22} \\ \sigma_{33} \\ \sigma_{i<j} \end{bmatrix} = \frac{E}{(1+\nu)(1-2\nu)} \begin{bmatrix} 1-\nu & \nu & \nu & 0 \\ \nu & 1-\nu & \nu & 0 \\ \nu & \nu & 1-\nu & 0 \\ 0 & 0 & 0 & \dfrac{(1-2\nu)}{2} \end{bmatrix} \begin{bmatrix} \varepsilon_{11} \\ \varepsilon_{22} \\ \varepsilon_{33} \\ 2\varepsilon_{i<j} \end{bmatrix} \tag{2.94}$$

$$\begin{bmatrix} \varepsilon_{11} \\ \varepsilon_{22} \\ \varepsilon_{33} \\ 2\varepsilon_{i<j} \end{bmatrix} = \frac{1}{E} \begin{bmatrix} 1 & -\nu & -\nu & 0 \\ -\nu & 1 & -\nu & 0 \\ -\nu & -\nu & 1 & 0 \\ 0 & 0 & 0 & 2(1+\nu) \end{bmatrix} \begin{bmatrix} \sigma_{11} \\ \sigma_{22} \\ \sigma_{33} \\ \sigma_{i<j} \end{bmatrix} \tag{2.95}$$

Table 2.1 provides the conversions between the coefficients of linear elasticity.

Table 2.1. Coefficients of linear elasticity

E	ν	λ	μ
$\frac{\mu(3\lambda+2\mu)}{\lambda+\mu}$	$\frac{\lambda}{2(\lambda+\mu)}$	$\frac{\nu E}{(1+\nu)(1-2\nu)}$	$\frac{E}{2(1+\nu)}$

2.2.5 Anisotropic Elasticity

In the case of linear *non-isotropic* materials, the stiffness matrix $\mathbf{K}^E$ of the generalized Hooke law (2.92) may not especially contain null coefficients and be symmetric. However, it may be expressed in some cases where the material has three mutually orthogonal planes of symmetry (Mal 91). Such material is said *orthotropic* and is represented by a constitutive relation such as:

$$\begin{bmatrix} \varepsilon_{11} \\ \varepsilon_{22} \\ \varepsilon_{33} \\ 2\varepsilon_{i<j} \end{bmatrix} = \begin{bmatrix} \frac{1}{E_1} & -\frac{\nu_{21}}{E_2} & -\frac{\nu_{31}}{E_3} & 0 \\ -\frac{\nu_{12}}{E_1} & \frac{1}{E_2} & -\frac{\nu_{32}}{E_3} & 0 \\ -\frac{\nu_{13}}{E_1} & -\frac{\nu_{23}}{E_2} & \frac{1}{E_3} & 0 \\ 0 & 0 & 0 & \frac{1}{G_{i<j}} \end{bmatrix} \begin{bmatrix} \sigma_{11} \\ \sigma_{22} \\ \sigma_{33} \\ \sigma_{i<j} \end{bmatrix} \quad \text{with} \quad \frac{\nu_{ij}}{E_i} = \frac{\nu_{ji}}{E_j} \tag{2.96}$$

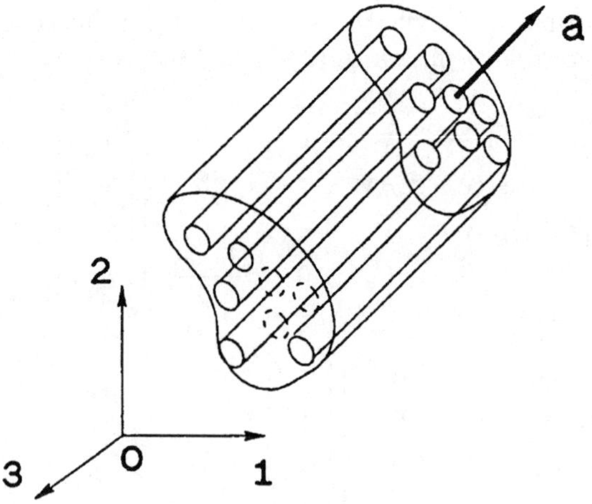

Fig. 2.4. Transverse isotropic material
(reprinted from (Mal 91) with permission of Prentice-Hall)

Transverse Isotropy. In the particular case where any plane containing a given axis is a plane of symmetry, as in unidirectional composite fibered materials, the material is said *transverse isotropic* (Fig. 2.4) and the previous relation reduces to the following one with only five unknown coefficients E_1, E_2, G_{12}, ν_{12}, ν:

$$\begin{bmatrix} \varepsilon_{11} \\ \varepsilon_{22} \\ \varepsilon_{33} \\ 2\varepsilon_{12} \\ 2\varepsilon_{13} \\ 2\varepsilon_{23} \end{bmatrix} = \begin{bmatrix} \frac{1}{E_1} & -\frac{\nu_{21}}{E_2} & -\frac{\nu_{21}}{E_2} & 0 & 0 & 0 \\ -\frac{\nu_{12}}{E_1} & \frac{1}{E_2} & -\frac{\nu}{E_2} & 0 & 0 & 0 \\ -\frac{\nu_{12}}{E_1} & -\frac{\nu}{E_2} & \frac{1}{E_2} & 0 & 0 & 0 \\ 0 & 0 & 0 & \frac{1}{G_{12}} & 0 & 0 \\ 0 & 0 & 0 & 0 & \frac{1}{G_{12}} & 0 \\ 0 & 0 & 0 & 0 & 0 & \frac{2(1+\nu)}{E_2} \end{bmatrix} \begin{bmatrix} \sigma_{11} \\ \sigma_{22} \\ \sigma_{33} \\ \sigma_{12} \\ \sigma_{13} \\ \sigma_{23} \end{bmatrix} \qquad \frac{\nu_{ij}}{E_i} = \frac{\nu_{ji}}{E_j} \tag{2.97}$$

2.3 Linear Viscoelasticity

2.3.1 Description

Fung described the viscoelastic properties as follows: *When a body is suddenly strained and then the strain is maintained constant afterward, the corresponding stresses induced in the body decrease with time: this phenomenon is called "stress relaxation". If the body is suddenly stressed and then the stress is maintained constant afterward, the body continues to deform: this phenomenon is called "creep". If the body is subjected to a cyclic loading, the stress-strain relationship in the loading process is usually somewhat different from that in the unloading process: this phenomenon is called "hysteresis". These behaviors are all features of a mechanical property called "viscoelasticity"* (Fung 93) (Fung 94).

Practical experience with materials which exhibit these behaviors suggests that they remember the previous deformations to which they have been subjected. This leads to formulate the general linear viscoelastic behavior as a relationship between the stress, the strain and their derivatives with respect to time in the form:

$$\left[\mathbf{a}_0 + \mathbf{a}_1\frac{\partial}{\partial t} + \mathbf{a}_2\frac{\partial^2}{\partial t^2} + \ldots\right]\boldsymbol{\sigma}(t) = \left[\boldsymbol{\alpha}_0 + \boldsymbol{\alpha}_1\frac{\partial}{\partial t} + \boldsymbol{\alpha}_2\frac{\partial^2}{\partial t^2} + \ldots\right]\boldsymbol{\epsilon}(t) \tag{2.98}$$

2.3.2 Solution for Stress Relaxation

The solution of (2.98) may be given in integral form as:

$$\boldsymbol{\sigma}(t) = \boldsymbol{\Phi}(t)\,\boldsymbol{\epsilon}(0) + \int_0^t \boldsymbol{\Phi}(t-\tau)\frac{\partial \boldsymbol{\epsilon}(\tau)}{\partial \tau}\,d\tau \tag{2.99}$$

where $\boldsymbol{\Phi}$ is the tensorial *relaxation* function of the material (Lockett 72). It appears in this relation that the stress $\boldsymbol{\sigma}$ depends on the strains $\boldsymbol{\epsilon}_\tau$ to which the material was subjected at all previous times τ up to time t.

Assuming no deformation in the initial state, (2.99) reduces to:

$$\boldsymbol{\sigma}(t) = \int_0^t \boldsymbol{\Phi}(t-\tau)\,\dot{\boldsymbol{\epsilon}}(\tau)\,d\tau \tag{2.100}$$

Isotropy. For isotropic materials, (2.100) may be developed into:

$$\boldsymbol{\sigma}(t) = \int_0^t \left[\, \lambda(t-\tau)\,\mathrm{Tr}(\dot{\boldsymbol{\epsilon}}(\tau))\,\mathbf{I} + 2\,\mu(t-\tau)\,\dot{\boldsymbol{\epsilon}}(\tau) \,\right] d\tau \qquad \text{i.e.:}$$

$$\boldsymbol{\sigma}(t) = \left[\, \int_0^t \lambda(t-\tau)\,\mathrm{Tr}(\dot{\boldsymbol{\epsilon}}(\tau))\,d\tau \,\right]\mathbf{I} + 2\int_0^t \mu(t-\tau)\,\dot{\boldsymbol{\epsilon}}(\tau)\,d\tau \tag{2.101}$$

which leads to common models such as the *Kelvin–Voigt* model (Lemaitre 90):

$$\boldsymbol{\sigma} = \lambda\left[\, \mathrm{Tr}(\boldsymbol{\epsilon}) + \theta_\lambda \mathrm{Tr}(\dot{\boldsymbol{\epsilon}}) \,\right]\mathrm{I} + 2\mu\left[\, \boldsymbol{\epsilon} + \theta_\mu \dot{\boldsymbol{\epsilon}} \,\right] \quad \text{which may be written as:} \tag{2.102}$$

$$\boldsymbol{\sigma} = \mathbf{K}^{\mathrm{E}}\boldsymbol{\epsilon} + \mathbf{D}^{\mathrm{V}}\dot{\boldsymbol{\epsilon}} \qquad \text{with} \qquad \begin{cases} \mathbf{K}^{\mathrm{E}}\boldsymbol{\epsilon} = \lambda\,\mathrm{Tr}(\boldsymbol{\epsilon})\,\mathrm{I} + 2\,\mu\,\boldsymbol{\epsilon} \\ \mathbf{D}^{\mathrm{V}}\dot{\boldsymbol{\epsilon}} = \lambda\,\theta_\lambda\,\mathrm{Tr}(\dot{\boldsymbol{\epsilon}})\,\mathrm{I} + 2\,\mu\,\theta_\mu\,\dot{\boldsymbol{\epsilon}} \end{cases} \tag{2.103}$$

where λ, μ are the Lamé coefficients defined in (2.91) for elasticity, and θ_λ, θ_μ are constants for viscosity. With $\boldsymbol{\sigma}$ and $\boldsymbol{\epsilon}$ as defined in (2.56), (2.103) becomes:

$$\boldsymbol{\sigma} = \boldsymbol{K}^{\mathrm{E}}\boldsymbol{\epsilon} + \boldsymbol{D}^{\mathrm{V}}\dot{\boldsymbol{\epsilon}} \qquad \text{with} \qquad \begin{cases} \boldsymbol{K}^{\mathrm{E}}\boldsymbol{\epsilon} = \lambda\,\mathrm{Tr}(\boldsymbol{\epsilon})\,\mathrm{I} + 2\,\mu\,\boldsymbol{\epsilon} \\ \boldsymbol{D}^{\mathrm{V}}\dot{\boldsymbol{\epsilon}} = \lambda\,\theta_\lambda\,\mathrm{Tr}(\dot{\boldsymbol{\epsilon}})\,\mathrm{I} + 2\,\mu\,\theta_\mu\,\dot{\boldsymbol{\epsilon}} \end{cases} \tag{2.104}$$

Similar expressions may be written for finite deformation, using the Lagrangian strain $\mathbf{E}$ and Piola–Kirchhoff second stress $\mathbf{S}$ tensors (Katona 78) (Pipkin 86).

2.3.3 Solution for Creep

The solution of (2.98) may be conversely given in the form:

$$\boldsymbol{\epsilon}(t) = \boldsymbol{\Psi}(t)\ \boldsymbol{\sigma}(0) + \int_0^t \boldsymbol{\Psi}(t-\tau)\ \dot{\boldsymbol{\sigma}}(\tau)\ d\tau \tag{2.105}$$

where $\boldsymbol{\Psi}$ is the tensorial *creep* function of the material (Lockett 72). Assuming no stress in the initial state, this reduces to:

$$\boldsymbol{\epsilon}(t) = \int_0^t \boldsymbol{\Psi}(t-\tau)\ \dot{\boldsymbol{\sigma}}(\tau)\ d\tau \tag{2.106}$$

Isotropy. For isotropic materials, (2.106) may be developed into:

$$\boldsymbol{\epsilon}(t) = \int_0^t \left[\ \chi(t-\tau)\ \mathrm{Tr}(\dot{\boldsymbol{\sigma}}(\tau))\ \mathrm{I} + 2\ \zeta(t-\tau)\ \dot{\boldsymbol{\sigma}}(\tau)\ \right] d\tau \qquad \text{i.e.:}$$

$$\boldsymbol{\epsilon}(t) = \left[\ \int_0^t \chi(t-\tau)\ \mathrm{Tr}(\dot{\boldsymbol{\sigma}}(\tau))\ d\tau\right] \mathrm{I} + 2 \int_0^t \zeta(t-\tau)\ \dot{\boldsymbol{\sigma}}(\tau)\ d\tau \tag{2.107}$$

which may lead to common models such as *Maxwell* model (Lemaitre 90):

$$\dot{\boldsymbol{\epsilon}} = \frac{1+\nu}{E}\left[\dot{\boldsymbol{\sigma}} + \frac{\boldsymbol{\sigma}}{\eta_1}\right] - \frac{\nu}{E}\left[\mathrm{Tr}(\dot{\boldsymbol{\sigma}}) + \frac{\mathrm{Tr}(\boldsymbol{\sigma})}{\eta_2}\right]\mathbf{I} \quad \text{which may be written as:} \tag{2.108}$$

$$\dot{\boldsymbol{\epsilon}} = \mathbf{H}^{E}\boldsymbol{\sigma} + \mathbf{N}^{V}\dot{\boldsymbol{\sigma}} \qquad \text{with} \qquad \begin{cases} \mathbf{H}^{E}\boldsymbol{\sigma} = \dfrac{1+\nu}{E\eta_1}\boldsymbol{\sigma} - \dfrac{\nu}{E\eta_2}\mathrm{Tr}(\boldsymbol{\sigma})\ \mathbf{I} \\ \mathbf{N}^{V}\dot{\boldsymbol{\sigma}} = \dfrac{1+\nu}{E\eta_1}\dot{\boldsymbol{\sigma}} - \dfrac{\nu}{E\eta_2}\mathrm{Tr}(\dot{\boldsymbol{\sigma}})\ \mathbf{I} \end{cases} \tag{2.109}$$

where E, ν are the Young modulus and Poisson coefficient (2.94), and η_1, η_2 are viscosity constants. With $\boldsymbol{\sigma}$ and $\boldsymbol{\epsilon}$ as defined in (2.56), (2.109) becomes:

$$\dot{\boldsymbol{\epsilon}} = \boldsymbol{H}^{E}\boldsymbol{\sigma} + \boldsymbol{N}^{V}\dot{\boldsymbol{\sigma}} \qquad \text{with} \qquad \begin{cases} \boldsymbol{H}^{E}\boldsymbol{\sigma} = \dfrac{1+\nu}{E\eta_1}\boldsymbol{\sigma} - \dfrac{\nu}{E\eta_2}\mathrm{Tr}(\boldsymbol{\sigma})\ \mathbf{I} \\ \boldsymbol{N}^{V}\dot{\boldsymbol{\sigma}} = \dfrac{1+\nu}{E\eta_1}\dot{\boldsymbol{\sigma}} - \dfrac{\nu}{E\eta_2}\mathrm{Tr}(\dot{\boldsymbol{\sigma}})\ \mathbf{I} \end{cases} \tag{2.110}$$

Similar expressions may be written for finite deformation, using the Lagrangian strain $\mathbf{E}$ and Piola–Kirchhoff second stress $\mathbf{S}$ tensors (Katona 78) (Pipkin 86).

2.3.4 Series Decomposition

The solutions for stress relaxation and creep have been given above in their general forms. Unless particular derivations like the Kelvin–Voigt (2.102) or Maxwell models (2.108) are used, the general isotropic relations (2.101) and (2.107) are not suitable in practice for solving viscoelastic problems. Another formulation must be used (Katona 80). The series decomposition method is here developed for stress relaxation but it may also be similarly developed to creep.

The stress relaxation solution was given in (2.99) as:

$$\boldsymbol{\sigma}(t) = \boldsymbol{\Phi}(t)\, \boldsymbol{\epsilon}(0) + \int_0^t \boldsymbol{\Phi}(t-\tau) \frac{\partial \boldsymbol{\epsilon}(\tau)}{\partial \tau}\, d\tau \tag{2.111}$$

In this one, the last integral term may be integrated by parts as:

$$\int_0^t \boldsymbol{\Phi}(t-\tau) \frac{\partial \boldsymbol{\epsilon}(\tau)}{\partial \tau}\, d\tau = \left[\, \boldsymbol{\Phi}(t-\tau)\, \boldsymbol{\epsilon}(\tau)\, \right]_0^t - \int_0^t \frac{\partial \boldsymbol{\Phi}(t-\tau)}{\partial \tau} \boldsymbol{\epsilon}(\tau)\, d\tau \tag{2.112}$$

to obtain after restitution in (2.111):

$$\boldsymbol{\sigma}(t) = \boldsymbol{\Phi}(0)\, \boldsymbol{\epsilon}(t) - \int_0^t \frac{\partial \boldsymbol{\Phi}(t-\tau)}{\partial \tau} \boldsymbol{\epsilon}(\tau)\, d\tau \tag{2.113}$$

Isotropy. For isotropic materials, the relaxation function $\boldsymbol{\Phi}$ may be decomposed into *bulk* and *shear* contributions:

$$\boldsymbol{\Phi}(t) = \mathbf{D}^P P(t) + \mathbf{D}^Q Q(t) \tag{2.114}$$

where $P(t)$ and $Q(t)$ are the *bulk* and *shear relaxation* functions, with:

$$\mathbf{D}^P = \begin{bmatrix} 1 & 1 & 1 & 0 & 0 & 0 \\ 1 & 1 & 1 & 0 & 0 & 0 \\ 1 & 1 & 1 & 0 & 0 & 0 \\ 0 & 0 & 0 & 0 & 0 & 0 \\ 0 & 0 & 0 & 0 & 0 & 0 \\ 0 & 0 & 0 & 0 & 0 & 0 \end{bmatrix} \quad \text{and} \quad \mathbf{D}^Q = \begin{bmatrix} \frac{4}{3} & -\frac{2}{3} & -\frac{2}{3} & 0 & 0 & 0 \\ -\frac{2}{3} & \frac{4}{3} & -\frac{2}{3} & 0 & 0 & 0 \\ -\frac{2}{3} & -\frac{2}{3} & \frac{4}{3} & 0 & 0 & 0 \\ 0 & 0 & 0 & 1 & 0 & 0 \\ 0 & 0 & 0 & 0 & 1 & 0 \\ 0 & 0 & 0 & 0 & 0 & 1 \end{bmatrix} \tag{2.115}$$

Assuming that the material response depends more on the recent history than on earlier events, $P(t)$ and $Q(t)$ should be *monotonically decreasing* functions as:

$$P(t) = P_\infty + \sum_{i=1}^{n_P} P_i\, e^{-\frac{t}{\beta_i}} \qquad \text{and} \qquad Q(t) = Q_\infty + \sum_{i=1}^{n_Q} Q_i\, e^{-\frac{t}{\gamma_i}} \tag{2.116}$$

where the P_i, Q_i and β_i, γ_i are the *relaxation moduli* and *relaxation times* for bulk and shear, respectively (Katona 80). After restitution in (2.113), comes:

$$\boldsymbol{\sigma}(t) = \boldsymbol{\Phi}(0)\,\boldsymbol{\epsilon}(t) - \int_0^t \left(\mathbf{D}^P \sum_{i=1}^{n_P} \frac{P_i}{\beta_i}\, e^{-\frac{t-\tau}{\beta_i}} + \mathbf{D}^Q \sum_{i=1}^{n_Q} \frac{Q_i}{\gamma_i}\, e^{-\frac{t-\tau}{\gamma_i}} \right) \boldsymbol{\epsilon}(\tau)\, d\tau \qquad \text{i.e.:}$$

$$\boldsymbol{\sigma}(t) = \mathbf{K}^E \boldsymbol{\epsilon}(t) - \mathbf{D}^P \sum_{i=1}^{n_P} P_i\, \mathbf{p}_i - \mathbf{D}^Q \sum_{i=1}^{n_Q} Q_i\, \mathbf{q}_i \qquad \text{with} \tag{2.117}$$

$$\mathbf{K}^E = \boldsymbol{\Phi}(0) = \mathbf{D}^P P(0) + \mathbf{D}^Q Q(0) \qquad \mathbf{p}_i = \int_0^t \frac{\boldsymbol{\epsilon}(\tau)}{\beta_i} e^{-\frac{t-\tau}{\beta_i}}\, d\tau \qquad \mathbf{q}_i = \int_0^t \frac{\boldsymbol{\epsilon}(\tau)}{\gamma_i} e^{-\frac{t-\tau}{\gamma_i}}\, d\tau$$

Using $\boldsymbol{\sigma}$ and $\boldsymbol{\epsilon}$ as defined in (2.56), (2.117) may be written in vectorial form as:

$$\boldsymbol{\sigma}(t) = \boldsymbol{K}^E \boldsymbol{\epsilon}(t) - \boldsymbol{D}^P \sum_{i=1}^{n_P} P_i\, \boldsymbol{p}_i - \boldsymbol{D}^Q \sum_{i=1}^{n_Q} Q_i\, \boldsymbol{q}_i \qquad \text{with} \tag{2.118}$$

$$\boldsymbol{K}^E = \boldsymbol{\Phi}(0) = \boldsymbol{D}^P P(0) + \boldsymbol{D}^Q Q(0) \qquad \boldsymbol{p}_i = \int_0^t \frac{\boldsymbol{\epsilon}(\tau)}{\beta_i} e^{-\frac{t-\tau}{\beta_i}}\, d\tau \qquad \boldsymbol{q}_i = \int_0^t \frac{\boldsymbol{\epsilon}(\tau)}{\gamma_i} e^{-\frac{t-\tau}{\gamma_i}}\, d\tau$$

where $\boldsymbol{D}^P$, $\boldsymbol{D}^Q$ are matrix adjusted from $\mathbf{D}^P$, $\mathbf{D}^Q$ to fit the vectorial forms. Similar expressions may be written for finite deformation, using the Lagrangian strain $\mathbf{E}$ and the Kirchhoff second stress $\mathbf{S}$ tensors (Katona 78) (Pipkin 86):

$$\mathbf{S}(t) = \mathbf{K}^E \mathbf{E}(t) - \mathbf{D}^P \sum_{i=1}^{n_P} P_i\, \mathbf{P}_i - \mathbf{D}^Q \sum_{i=1}^{n_Q} Q_i\, \mathbf{Q}_i \qquad \text{with} \tag{2.119}$$

$$\mathbf{K}^E = \boldsymbol{\Phi}(0) = \mathbf{D}^P P(0) + \mathbf{D}^Q Q(0) \qquad \mathbf{P}_i = \int_0^t \frac{\mathbf{E}(\tau)}{\beta_i} e^{-\frac{t-\tau}{\beta_i}}\, d\tau \qquad \mathbf{Q}_i = \int_0^t \frac{\mathbf{E}(\tau)}{\gamma_i} e^{-\frac{t-\tau}{\gamma_i}}\, d\tau$$

With (2.117) or (2.119), relations may be derived for incremental resolution.

Conclusion

The theoretical relationships presented in this chapter are purely mathematical expressions. They are based on descriptions with respect to coordinate systems following the deformation of the material: if the deformation is small, every measure may be done with respect to the known resting state of the material, whereas if the deformation is large, substitute equivalent variables must be considered to go ahead in the analysis with the known reference configuration. These variables are the Lagrangian strain and Kirchhoff stress tensors $\mathbf{E}$ and $\mathbf{S}$. In any case, if the material is non-linear, physical non-linearities will appear in its constitutive relation within the coefficients and the quadratic tensorial terms of its stress-strain relationship. Dealing with non-linearities, mathematical expressions are not constant over space and time. Solutions can then only be reached step-by-step by the way of incremental iterative methods, which provide only approximate solutions to a limited number of problems. The principles of these numerical methods are presented in the following chapter, with a particular insight to the finite element method commonly used in mechanical analyses.

References

This chapter is essentially based on (Ciarlet 86), (Leipholz 74), (Trompette 92), (Mal 91), (Oden 89), (Ogden 84), (Katona 78, 80), (Lockett 72), (CHARM D4). For deeper investigation, refer also to (Lemaitre 90), (Marsden 83), (Naghdi 94), (Timoshenko 70), (Christensen 82), (Coleman 61), (Zienkiewicz 91).

Allaire 77	P.E. Allaire, J.G. Thacker, R.F. Edlich, G.J. Rodenheaver, M.T. Edgerton (1977), Finite deformation theory for in-vivo human skin, *J. Bioeng.*, 1, 239–249
Christensen 82	R.M. Christensen (1982), *Theory of viscoelasticity: an introduction*, 2nd edn. New York: Academic Press
CHARM D4	W. Maurel, Y. Wu (1994), Survey of mechanical models for tissue deformation and muscle contraction with identification of parametric data, LIG-EPFL/MIRALab-UG, ESPRIT 9036 Project CHARM, Internal Deliverable D4
Ciarlet 86	P.G. Ciarlet (1986), *Elasticité tridimensionnelle*. Paris: Masson
Coleman 61	B.D. Coleman, W. Noll (1961), Foundations of Linear Viscoelasticity, *Review of Modern Physics*, 33, 2, 239–249
Fung 93	Y.C. Fung (1993), *Biomechanics: Motion, Flow, Stress and Growth*. New York: Springer-Verlag

Fung 94	Y.C. Fung (1994), *A First Course in Continuum Mechanics: for Physical and Biological Engineers and Scientists*, 3rd edn., Englewood Cliffs, NJ: Prentice Hall
Katona 78	M.G. Katona (1978), A viscoelastic-plastic constitutive model with a finite element solution methodology: technical final report, 1974–1977; prep.: Civil Engineering Laboratory, Naval Construction Battalion Center, Port Hueneme, California 93043
Katona 80	M.G. Katona (1980), Combo-viscoplasticity: an introduction with incremental formulation, *Computers & Structures*, 11, 217–224
Leipholz 74	H. Leipholz (1974), *Theory of Elasticity*. Leyden: Nordhoff
Lemaitre 90	J. Lemaitre, J.-L. Chaboche (1990), *Mechanics of solid materials* (originally published in French as Mécanique des matériaux solides by Dunod, Paris, 1985 and cop. Bordas, Paris, 1985) transl. by B. Shrivastava. Cambridge: Cambridge University Press
Lockett 72	F.J. Lockett (1972), *Non-linear Viscoelastic Solids*, London: Academic Press
Mal 91	A.K. Mal, S.J. Singh (1991), *Deformation of elastic solids*, Englewood Cliffs, NJ: Prentice-Hall
Marsden 83	J.E. Marsden, T.J.R. Hughes (1983), *Mathematical foundations of Elasticity*, Englewood Cliffs, NJ: Prentice-Hall
Naghdi 94	P.M. Naghdi, A.J.M. Spencer, A.H. England (1994), *Non-linear Elasticity and Theoretical Mechanics: In Honour of A.E. Green*. Oxford: Oxford University Press
Oden 89	J.T. Oden, J.M. Steele (1989), *Finite Elements of Non-linear Continua*. New York: Dekker
Ogden 84	R.W. Ogden (1984), *Non-linear Elastic Deformations*. Chichester: Ellis Horwood/New York: John Wiley & Sons
Pipkin 86	A.C. Pipkin (1986), Large Deformations with Small Strains, in *Lectures on Viscoelasticity Theory*, *App. Math. Sci.*, 7, 115–128
Timoshenko 70	S.P. Timoshenko, J.N. Goodier (1970), *Theory of Elasticity*, 3rd edn. Auckland, NZ: McGraw-Hill
Trompette 92	P. Trompette (1992), *Mécanique des structures par la méthode des éléments finis: Statique et Dynamique*. Paris: Masson
Zienkiewicz 91	O.C. Zienkiewicz, R.L. Taylor (1991), *Solid and Fluid Mechanics, Dynamics and Non-linearity*, 4th edn. London: McGraw-Hill

3 Resolution Methods

Theoretical mechanics, as presented in Chapter 2, only provides differential equations describing the properties and behavior of materials. Resolution methods are necessary to obtain temporal descriptions of the state variables as required for the simulation. The assumption of complete linearity is a great advantage because it allows the solution of the complete dynamic problem, and leads to the exact response of the system. However, for a physically realistic simulation, all the non-linearities of a material deformation must be taken into account. In this case, solutions can only be approximated using incremental-iterative procedures after spatial and temporal discretizations. The common approach for this purpose is to combine a finite element discretization of the geometry together with a finite difference discretization of time and an updated Lagrangian iterative scheme based on the Newton–Raphson method. Basics of these methods are presented in this chapter for helping to understand how complex non-linear deformation processes and material constitutive relations may be applied to soft tissue simulation.

3.1 The Finite Element Method

3.1.1 Principle

The basic idea of the finite element method is to approximate the continuous domain Ω to a mesh of smaller finite elements with simple regular shape. Thus, any point P of the domain Ω may be expressed within an element, by interpolation between the surrounding mesh nodes N_i, using of interpolation functions named *shape functions*. Then, any unknown function g(P) applied to Ω may also be approximated as a function of the interpolation mesh.

As a result of the method, the unknown continuous problem $g(P)$ is approximated by a finite system of equations $g(N_i)$ in terms of node coordinates. The precision of the approximation improves with the level of discretization and the degree of interpolation chosen, while to the detriment of the necessary computation time and memory size (Zienkiewicz 89).

3.1.2 Geometric Discretization

The general elementary interpolation relation may be written in the form:

$$\mathbf{X} = \mathbf{H}\,\mathbf{N} \tag{3.1}$$

where **H** is the tensorial shape-function
N the vector of the elementary nodal coordinates
X the coordinate vector of any point in the element

In general, finite element meshes are composed of elements with simple geometric shapes such as segments for curves, triangles or quadrilaterals for surfaces (Fig. 3.1) (Trompette 92), and tetrahedrons or hexahedrons for volumes (Fig. 3.3). However, any other shape may be considered as well for creating the mesh, as long as it is compatible with the dimension of the space to discretize. In practice, finite element softwares provide huge libraries of elements, not only distinct in shape or interpolation degree but also in the properties and functions assigned to them, depending on the kind of analysis to be performed.

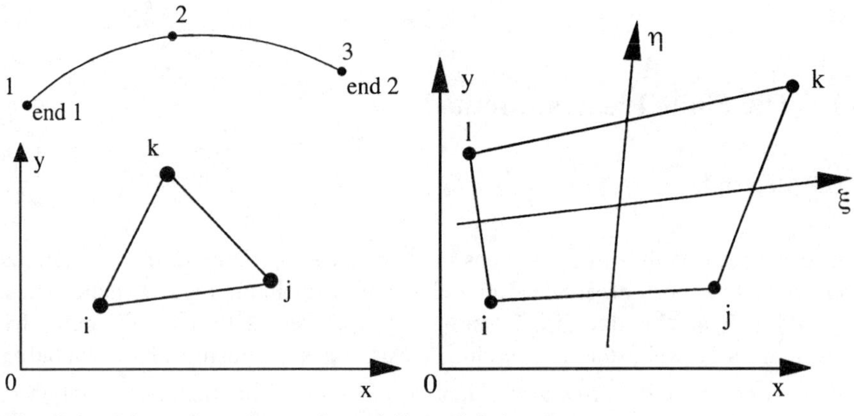

Fig. 3.1. Line, triangular, and quadrilateral finite elements
(reprinted from (Trompette 92) with permission of Masson)

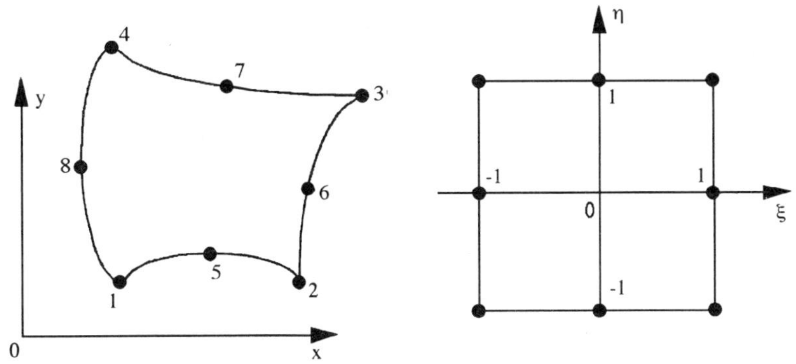

Fig. 3.2. Real and parent spaces of a quadrilateral finite element (reprinted from (Trompette 92) with permission of Masson)

Isoparametric Interpolations. There is a major interest in using simple geometric shapes because of their symmetric properties, which allow the use of reduced variables in the expression of the shape functions (Zienkiewicz 89). Considering for example a curve divided into segments, **H** may be chosen within each segment as function of a reduced parameter α, such that:

$$\mathbf{X} = \left(\frac{1-\alpha}{2}\right)\mathbf{N}_i + \left(\frac{1+\alpha}{2}\right)\mathbf{N}_{i+1} = \begin{bmatrix}\frac{1-\alpha}{2} & \frac{1+\alpha}{2}\end{bmatrix}\begin{bmatrix}\mathbf{N}_i \\ \mathbf{N}_{i+1}\end{bmatrix} \qquad t \in [0,1] \tag{3.2}$$

This expression corresponds to a linear interpolation within a line finite element. Such an expression moves the interpolation procedure from the real domain Ω to a parallel regular space named *parent space* of the element (Fig. 3.2). The shape functions may thus be taken identical for each element of the mesh. This helps for avoiding the repetition of calculations on different domains which can be approximated on the same polynomial basis. Such interpolations and their associated elements are then said *isoparametric*. Some common isoparametric elements are described hereafter with their shape functions (Trompette 92).

Lines. Curvilinear segments may be used for one-dimensional analysis:

$$\mathbf{X} = \begin{bmatrix}\frac{\xi(\xi-1)}{2} & \frac{\xi(\xi+1)}{2} & 1-\xi^2\end{bmatrix}\begin{bmatrix}\mathbf{N}_{i-1} \\ \mathbf{N}_i \\ \mathbf{N}_{i+1}\end{bmatrix} \qquad \text{quadratic interpolation} \tag{3.3}$$

In this case of quadratic (non-linear) interpolation, the element has two geometric ends but three interpolation nodes.

Triangles. Triangles may be used for two-dimensional analysis:

$$\mathbf{X} = [\xi \quad \eta \quad 1-\xi-\eta] \begin{bmatrix} \mathbf{N}_i \\ \mathbf{N}_j \\ \mathbf{N}_k \end{bmatrix} \qquad \text{linear interpolation} \tag{3.4}$$

$$\mathbf{X} = \left[t(2t-1) \quad \xi(2\xi-1) \quad \eta(2\eta-1) \quad 4\xi t \quad 4\xi\eta \quad 4\eta t \right] \begin{bmatrix} \mathbf{N}_i \\ \mathbf{N}_j \\ \mathbf{N}_k \\ \mathbf{N}_l \\ \mathbf{N}_m \\ \mathbf{N}_n \end{bmatrix} \qquad t = 1-\xi-\eta \qquad \text{6 - node - interpolation} \tag{3.5}$$

Quadrilaterals. Quadrilaterals may be used for two-dimensional analysis:

$$\mathbf{X} = \left[\frac{(1-\xi)(1-\eta)}{4} \quad \frac{(1+\xi)(1-\eta)}{4} \quad \frac{(1+\xi)(1+\eta)}{4} \quad \frac{(1-\xi)(1+\eta)}{4} \right] \begin{bmatrix} \mathbf{N}_i \\ \mathbf{N}_j \\ \mathbf{N}_k \\ \mathbf{N}_l \end{bmatrix} \tag{3.6}$$

linear interpolation

Tetrahedrons. Tetrahedrons may be used for three-dimensional analysis:

$$\mathbf{X} = [\alpha \quad \xi \quad \eta \quad \chi] \begin{bmatrix} \mathbf{N}_i \\ \mathbf{N}_j \\ \mathbf{N}_k \\ \mathbf{N}_l \end{bmatrix} \qquad \alpha = 1-\xi-\eta-\chi \qquad \text{linear interpolation} \tag{3.7}$$

Hexahedrons. Hexahedrons may be used for three-dimensional analysis:

$$\mathbf{X} = [h_i]^T [\mathbf{N}_i] \qquad \begin{cases} h_i = \frac{1}{8}(1 \pm \xi)\,(1 \pm \eta)\,(1 \pm \chi) \\ i = 1, 2, .., 8 \end{cases} \qquad \text{linear interpolation} \tag{3.8}$$

It is one of the elements the most frequently used in 3D modeling.

Tetrahedron and hexahedron elements are shown in Fig. 3.3.

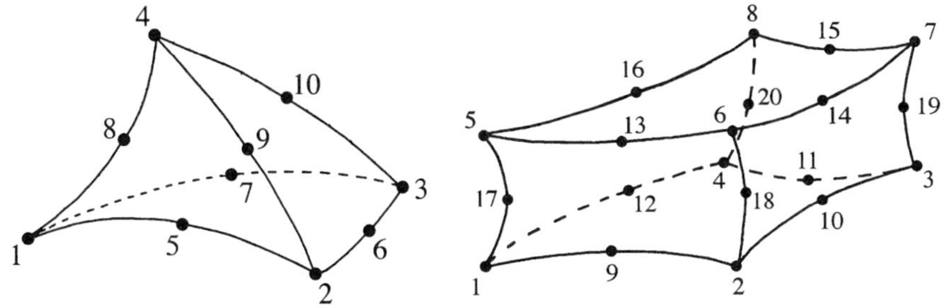

Fig. 3.3. Tetrahedron and hexahedron finite elements
(reprinted from Ansys User's Manual)

3.1.3 Application to Continuum Mechanics

Finite Deformations. The general equation of motion of the continuous medium has been provided in the Lagrangian form of the virtual work principle as (2.52):

$$\delta\hat{W}^{acc} = \delta\hat{W}^{ext} + \delta\hat{W}^{int} \tag{3.9}$$

with $\delta\hat{W}^{acc} = \int_V \delta\hat{\mathbf{u}}^T \rho_0 \mathbf{\Gamma}^V \, dV$ inertial virtual work

$\delta\hat{W}^{int} = -\int_V \mathrm{Tr}\left(\delta\hat{\mathbf{E}}^T\mathbf{S}\right) dV = -\int_V \delta\hat{\boldsymbol{E}}^T \boldsymbol{S} \, dV$ internal virtual work

$\delta\hat{W}^{ext} = \int_V \delta\hat{\mathbf{u}}^T \mathbf{f}^V \, dV + \int_S \delta\hat{\mathbf{u}}^T \mathbf{t}^S \mathbf{N} \, dS$ external virtual work

Assuming a finite element discretization as defined in (3.1), the displacement function $\mathbf{u}$ of any point P and its derivatives with respect to time may also be expressed in terms of the nodal displacement vector $\mathbf{U}$ as:

$$\mathbf{u} = \mathbf{H}\,\mathbf{U} \qquad \dot{\mathbf{u}} = \mathbf{H}\,\dot{\mathbf{U}} \qquad \ddot{\mathbf{u}} = \mathbf{H}\,\ddot{\mathbf{U}} \tag{3.10}$$

In particular, the displacements of the application points of the external body and contact forces $\mathbf{f}_n^V$ and $\mathbf{t}_n^S$ may be related to the nodal displacement vector $\mathbf{U}$ by:

$$\mathbf{u}^f = \mathbf{H}^f\mathbf{U} \qquad \mathbf{u}^t = \mathbf{H}^t\mathbf{U} \tag{3.11}$$

and the strain-displacement relationship may be expressed in terms $\mathbf{U}$ in the form:

$$\mathbf{E} = \mathbf{B}\,\mathbf{U} \tag{3.12}$$

where $\mathbf{B} \equiv \mathbf{B}(\mathbf{U})$ is a non-linear matricial differential operator obtained from (2.24). Then, assuming for the general case a non-linear viscoelastic constitutive relationship in the form $\mathbf{S}(\mathbf{E}, \dot{\mathbf{E}})$ the virtual work equation may be written as:

$$\delta\hat{W}^{acc} = \delta\hat{W}^{ext} + \delta\hat{W}^{int} \tag{3.13}$$

with: $\delta\hat{W}^{acc} = \int_V \delta\hat{\mathbf{u}}^T \rho_0 \mathbf{\Gamma}^V \, dV = \delta\hat{\mathbf{U}}^T \left(\int_V \rho_0 \mathbf{H}^T \mathbf{H} \, dV \right) \ddot{\mathbf{U}}$

$$\delta\hat{W}^{int} = -\int_V \delta\hat{\boldsymbol{E}}^T \boldsymbol{S} \, dV = -\delta\hat{\mathbf{U}}^T \left[\int_V \boldsymbol{B}^T \boldsymbol{S}(\mathbf{U}, \dot{\mathbf{U}}) \, dV \right]$$

$$\delta\hat{W}^{ext} = \int_V \delta\hat{\mathbf{u}}^{f^T} \mathbf{f}^V \, dV + \int_S \delta\hat{\mathbf{u}}^{t^T} \mathbf{t}^S \mathbf{N} \, dS = \delta\hat{\mathbf{U}}^T \left(\int_V \mathbf{H}^{f^T} \mathbf{f}^V \, dV + \int_S \mathbf{H}^{t^T} \mathbf{t}^S \mathbf{N} \, dS \right)$$

i.e., after removing the virtual nodal displacement $\delta\hat{\mathbf{U}}$

$$\mathbf{M}\,\ddot{\mathbf{U}} + \mathbf{\Pi}(\mathbf{U}, \dot{\mathbf{U}}) = \mathbf{L} \tag{3.14}$$

with: $\mathbf{M} = \int_V \rho_0 \mathbf{H}^T \mathbf{H} \, dV$ — *elementary* mass matrix

$\mathbf{\Pi}(\mathbf{U}, \dot{\mathbf{U}}) = \int_V \boldsymbol{B}^T \boldsymbol{S}(\mathbf{U}, \dot{\mathbf{U}}) \, dV$ — *elementary* internal force vector

$\mathbf{L} = \int_V \mathbf{H}^{f^T} \mathbf{f}^V \, dV + \int_S \mathbf{H}^{t^T} \mathbf{t}^S \mathbf{N} \, dS$ — *elementary* nodal external force vector

or after combining all the elementary equations into a global tensorial equation:

$$\underline{\mathbf{M}}\,\underline{\ddot{\mathbf{U}}} + \underline{\mathbf{\Pi}}(\underline{\mathbf{U}}, \underline{\dot{\mathbf{U}}}) = \underline{\mathbf{L}} \tag{3.15}$$

with $\underline{\mathbf{M}}$, $\underline{\mathbf{U}}$, $\underline{\mathbf{\Pi}}$ and $\underline{\mathbf{L}}$ as global mass matrix, nodal displacement, internal and external force vectors with forms similar to those appearing in (3.14) (Curnier 94).

As a result, the dynamic response of the mesh under the defined external loads may be analyzed by solving the finite second-order differential system for the global nodal displacements vector $\underline{\mathbf{U}}$, using integration methods depending on the linearity of the problem (§3.3). When the dynamic response of the system is not required, it has proved efficient to assume its evolution as *quasi-static*, i.e., a succession of equilibrium states, in order to remove from (3.15) the inertial and viscous terms due to acceleration and velocity, and get:

$$\underline{\mathbf{\Pi}}(\underline{\mathrm{U}}) = \underline{\mathbf{L}} \qquad (3.16)$$

In the following, equations for the elementary or global levels are indifferently designed *without underlining* since they are similar.

Linear Dynamics. In the case of small deformations (geometric linearity), the matrix and vectors defined in (3.14) may be computed in the reference configuration using the infinitesimal strain tensor $\boldsymbol{\epsilon}$ and the Cauchy stress tensor $\boldsymbol{\sigma}$ (Sect. 2.1), but the physical non-linearities remain in the expression of $\mathbf{\Pi}$, giving:

$$\mathbf{M}\,\ddot{\mathbf{U}} + \mathbf{\Pi}\left(\mathbf{U},\dot{\mathbf{U}}\right) = \mathbf{L} \qquad (3.17)$$

with: $\mathbf{M} = \int_V \rho_0 \mathbf{H}^T\mathbf{H}\,\mathrm{dV}$ *elementary* mass matrix

$\mathbf{\Pi}\left(\mathbf{U},\dot{\mathbf{U}}\right) = \int_V \boldsymbol{B}^T\boldsymbol{\sigma}\left(\mathbf{U},\dot{\mathbf{U}}\right)\mathrm{dV}$ *elementary* internal force vector

$\mathbf{L} = \int_V \mathbf{H}^{f^T}\mathbf{f}^V\,\mathrm{dV} + \int_S \mathbf{H}^{t^T}\mathbf{t}^S\mathbf{N}\,\mathrm{dS}$ *elementary* nodal external force vector

In case of a linear viscoelastic constitutive relationship $\boldsymbol{\sigma}(\boldsymbol{\epsilon},\dot{\boldsymbol{\epsilon}})$ such as the *Kelvin–Voigt* model (2.103) i.e., with uncoupled viscous and elastic components:

$\boldsymbol{\sigma} = \boldsymbol{K}^E\boldsymbol{\epsilon} + \boldsymbol{D}^V\dot{\boldsymbol{\epsilon}}$ using vectorial forms $\boldsymbol{\sigma}$ and $\boldsymbol{\epsilon}$ (2.56), (3.18)

where $\boldsymbol{K}^E$ and $\boldsymbol{D}^V$ are the constant stiffness and damping matrix of the material (Hughes 87), the elementary equation dynamics may then be written as usual:

$$\mathbf{M}\ddot{\mathbf{U}} + \mathbf{D}\dot{\mathbf{U}} + \mathbf{K}\mathbf{U} = \mathbf{L} \qquad (3.19)$$

with: $\mathbf{M} = \int_V \rho_0 \mathbf{H}^T\mathbf{H}\,\mathrm{dV}$ *elementary* mass matrix

$\mathbf{D} = \int_V \boldsymbol{B}^T\boldsymbol{D}^V\boldsymbol{B}\,\mathrm{dV}$ *elementary* damping matrix

$\mathbf{K} = \int_V \boldsymbol{B}^T\boldsymbol{K}^E\boldsymbol{B}\,\mathrm{dV}$ *elementary* stiffness matrix

$\mathbf{L} = \int_V \mathbf{H}^{f^T}\mathbf{f}^V\,\mathrm{dV} + \int_S \mathbf{H}^{t^T}\mathbf{t}^S\mathbf{N}\,\mathrm{dS}$ *elementary* nodal external force vector

3.1.4 2D Linear Dynamics Example

This section illustrates on a planar analysis how finite elements are applied in usual linear mechanical problems (Hughes 87). Let's consider a linear *Kelvin–Voigt* solid (2.103) undergoing small planar deformations when subjected to external forces. The first step is to select a basic finite element shape and to apply the meshing procedure accounting for specified points and dimensions in the mesh. Assuming a linear quadrilateral-based finite element discretization as:

$$\mathbf{X} = \mathbf{H}\,\mathbf{N} \qquad \mathbf{H} = \begin{bmatrix} \mathbf{h}_1 & \mathbf{h}_2 & \mathbf{h}_3 & \mathbf{h}_4 \end{bmatrix} \tag{3.20}$$

$$\mathbf{N} = \begin{bmatrix} \mathbf{N}_i^T & \mathbf{N}_j^T & \mathbf{N}_k^T & \mathbf{N}_l^T \end{bmatrix}^T = \begin{bmatrix} x_{1i} & x_{2i} & x_{1j} & x_{2j} & x_{1k} & x_{2k} & x_{1l} & x_{2l} \end{bmatrix}^T$$

$$h_1 = \frac{(1-\xi)(1-\eta)}{4}\begin{bmatrix} 1 & 0 \\ 0 & 1 \end{bmatrix} \qquad h_2 = \frac{(1+\xi)(1-\eta)}{4}\begin{bmatrix} 1 & 0 \\ 0 & 1 \end{bmatrix}$$

$$h_3 = \frac{(1+\xi)(1+\eta)}{4}\begin{bmatrix} 1 & 0 \\ 0 & 1 \end{bmatrix} \qquad h_4 = \frac{(1-\xi)(1+\eta)}{4}\begin{bmatrix} 1 & 0 \\ 0 & 1 \end{bmatrix}$$

the displacement $\mathbf{u} = \begin{bmatrix} \mathbf{u}_1 & \mathbf{u}_2 \end{bmatrix}^T$ of any point P and its derivatives with respect to time may also be expressed in terms of the nodal displacement vector $\mathbf{U}$ as:

$$\mathbf{u} = \mathbf{H}\,\mathbf{U} \qquad \dot{\mathbf{u}} = \mathbf{H}\,\dot{\mathbf{U}} \qquad \ddot{\mathbf{u}} = \mathbf{H}\,\ddot{\mathbf{U}} \tag{3.21}$$

In particular, the displacements of the application points of the external volume and surface forces $\mathbf{f}_n^V$ and $\mathbf{t}_n^S$ may be related to the nodal displacement vector $\mathbf{U}$:

$$\mathbf{u}^f = \mathbf{H}^f\mathbf{U} \qquad \mathbf{u}^t = \mathbf{H}^t\mathbf{U} \tag{3.22}$$

Then, the linear strain-displacement relationship defined in (2.28) takes the form:

$$\boldsymbol{\epsilon} = \boldsymbol{B}\,\mathbf{U} \quad \text{with} \quad \mathbf{U} = \begin{bmatrix} \mathbf{U}_i^T & \mathbf{U}_j^T & \mathbf{U}_k^T & \mathbf{U}_l^T \end{bmatrix}^T \tag{3.23}$$

$$\boldsymbol{\epsilon} = \begin{bmatrix} \varepsilon_{11} \\ \varepsilon_{22} \\ 2\varepsilon_{12} \end{bmatrix} \qquad \varepsilon_{11} = \frac{\partial u_1}{\partial X_1} \qquad \varepsilon_{22} = \frac{\partial u_2}{\partial X_2} \qquad \varepsilon_{12} = \frac{1}{2}\left(\frac{\partial u_1}{\partial X_2} + \frac{\partial u_2}{\partial X_1}\right) \tag{3.24}$$

$$\begin{bmatrix}\dfrac{\partial}{\partial X_1}\\ \dfrac{\partial}{\partial X_2}\end{bmatrix}=\begin{bmatrix}\dfrac{\partial X_1}{\partial\xi} & \dfrac{\partial X_2}{\partial\xi}\\ \dfrac{\partial X_1}{\partial\eta} & \dfrac{\partial X_2}{\partial\eta}\end{bmatrix}^{-1}\begin{bmatrix}\dfrac{\partial}{\partial\xi}\\ \dfrac{\partial}{\partial\eta}\end{bmatrix}\qquad\begin{bmatrix}\dfrac{\partial\mathbf{u}}{\partial\xi}\\ \dfrac{\partial\mathbf{u}}{\partial\eta}\end{bmatrix}=\begin{bmatrix}\dfrac{\partial\mathbf{h}_1}{\partial\xi} & \dfrac{\partial\mathbf{h}_2}{\partial\xi} & \dfrac{\partial\mathbf{h}_3}{\partial\xi} & \dfrac{\partial\mathbf{h}_4}{\partial\xi}\\ \dfrac{\partial\mathbf{h}_1}{\partial\eta} & \dfrac{\partial\mathbf{h}_2}{\partial\eta} & \dfrac{\partial\mathbf{h}_3}{\partial\eta} & \dfrac{\partial\mathbf{h}_4}{\partial\eta}\end{bmatrix}\mathbf{U}\tag{3.25}$$

and using a linear *Kelvin–Voigt* viscoelastic constitutive relationship in the form:

$$\boldsymbol{\sigma}=\boldsymbol{K}^{\mathrm{E}}\boldsymbol{\epsilon}+\boldsymbol{D}^{\mathrm{V}}\dot{\boldsymbol{\epsilon}}\qquad(2.103)\tag{3.26}$$

the linear variational formulation defined in (3.13) becomes:

$$\delta\hat{\mathrm{W}}^{\mathrm{acc}}=\delta\hat{\mathrm{W}}^{\mathrm{ext}}+\delta\hat{\mathrm{W}}^{\mathrm{int}}\tag{3.27}$$

with $\delta\hat{\mathrm{W}}^{\mathrm{acc}}=\int_{\mathrm{V}}\delta\hat{\mathbf{u}}^{\mathrm{T}}\rho_0\boldsymbol{\Gamma}^{\mathrm{V}}\,\mathrm{dV}=\delta\hat{\mathrm{U}}^{\mathrm{T}}\left(\int_{\mathrm{V}}\rho_0\mathbf{H}^{\mathrm{T}}\mathbf{H}\,\mathrm{dV}\right)\ddot{\mathrm{U}}$

$$\delta\hat{\mathrm{W}}^{\mathrm{int}}=-\int_{\mathrm{V}}\delta\hat{\boldsymbol{\epsilon}}^{\mathrm{T}}\boldsymbol{\sigma}\,\mathrm{dV}=-\delta\hat{\mathrm{U}}^{\mathrm{T}}\left(\boldsymbol{B}^{\mathrm{T}}\boldsymbol{K}^{\mathrm{E}}\boldsymbol{B}\,\mathrm{U}+\boldsymbol{B}^{\mathrm{T}}\boldsymbol{D}^{\mathrm{V}}\boldsymbol{B}\,\dot{\mathrm{U}}\right)$$

$$\delta\hat{\mathrm{W}}^{\mathrm{ext}}=\int_{\mathrm{V}}\delta\hat{\mathbf{u}}^{\mathrm{f}^{\mathrm{T}}}\mathbf{f}^{\mathrm{V}}\,\mathrm{dV}+\int_{\mathrm{S}}\delta\hat{\mathbf{u}}^{\mathrm{t}^{\mathrm{T}}}\mathbf{t}^{\mathrm{S}}\mathbf{N}\,\mathrm{dS}=\delta\hat{\mathrm{U}}^{\mathrm{T}}\left(\int_{\mathrm{V}}\mathbf{H}^{\mathrm{f}^{\mathrm{T}}}\mathbf{f}^{\mathrm{V}}\,\mathrm{dV}+\int_{\mathrm{S}}\mathbf{H}^{\mathrm{t}^{\mathrm{T}}}\mathbf{t}^{\mathrm{S}}\mathbf{N}\,\mathrm{dS}\right)$$

i.e., after removing the virtual nodal displacement $\delta\hat{\mathrm{U}}$:

$$\mathbf{M}\,\ddot{\mathrm{U}}+\mathbf{D}\,\dot{\mathrm{U}}+\mathbf{K}\,\mathrm{U}=\mathbf{L}\tag{3.28}$$

with: $\mathbf{M}=\int_{\mathrm{V}}\rho_0\mathbf{H}^{\mathrm{T}}\mathbf{H}\,\mathrm{dV}$ — *elementary* mass matrix

$\mathbf{D}=\int_{\mathrm{V}}\boldsymbol{B}^{\mathrm{T}}\boldsymbol{D}^{\mathrm{V}}\boldsymbol{B}\,\mathrm{dV}$ — *elementary* damping matrix

$\mathbf{K}=\int_{\mathrm{V}}\boldsymbol{B}^{\mathrm{T}}\boldsymbol{K}^{\mathrm{E}}\boldsymbol{B}\,\mathrm{dV}$ — *elementary* stiffness matrix

$\mathbf{L}=\int_{\mathrm{V}}\mathbf{H}^{\mathrm{f}^{\mathrm{T}}}\mathbf{f}^{\mathrm{V}}\,\mathrm{dV}+\int_{\mathrm{S}}\mathbf{H}^{\mathrm{t}^{\mathrm{T}}}\mathbf{t}^{\mathrm{S}}\mathbf{N}\,\mathrm{dS}$ — *elementary* nodal external force vector

Finally, the dynamics of the mesh for defined external loads may be analyzed by solving the linear second-order differential system for **U** using linear integration algorithms (Hughes 87). In general, the equation dynamics cannot be obtained in the common linear form given by (3.28). As shown hereafter, incremental variables must be used in order to approximate the non-linear time-continuous system into piecewise linear relations allowing an incremental resolution.

3.2 Incremental Description

3.2.1 Incremental Variables

Dealing with non-linearities, solutions can only be approximated step-by-step by means of incremental / iterative methods. For this purpose, all relations must be converted into their incremental / iterative forms. As a result of the formulation, the non-linear equations are replaced by linear incremental / iterative relations in which the tensorial terms must be updated at each increment / iteration step. Considering finite deformations, two approaches may be applied: the *Lagrangian* approach and the *Eulerian* approach (Argyris 80). The Eulerian approach refers all the kinematic and state variables to the current configuration, while the Lagrangian approach refers to the previous configurations of the continuum (Fig. 3.4) (Kleiber 89). If they are referred to the undeformed (original) configuration, the approach is called *the Total Lagrangian* approach, and if they are referred to the recently computed configuration, the approach is called *the Updated Lagrangian* or *Approximate Eulerian.* In practice, the Updated Lagrangian approach has proved to be more general and computationally efficient. The Eulerian approach is not convenient because the current configuration is the unknown of the problem. Furthermore, as it belongs to the current state, talking about a Cauchy stress increment has no correct meaning, because it would relate two stresses described in two different configurations. Any previous Lagrangian configuration provides then a convenient reference basis to describe the state variables, and the Piola–Kirchhoff stress and Green–Lagrange strain may be considered as suitable variables for the incremental formulation (Desai 80).

Within each step of the incremental process, the increments in stress $\mathbf{S}_n \equiv \mathbf{S}(t_n)$ and strain $\mathbf{E}_n \equiv \mathbf{E}(t_n)$ may be defined as:

$$\Delta\mathbf{S} = \mathbf{S}_{n+1} - \mathbf{S}_n \qquad \Delta\mathbf{E} = \mathbf{E}_{n+1} - \mathbf{E}_n \qquad \text{i.e.:} \tag{3.29}$$

$$\Delta\boldsymbol{S} = \boldsymbol{S}_{n+1} - \boldsymbol{S}_n \qquad \Delta\boldsymbol{E} = \boldsymbol{E}_{n+1} - \boldsymbol{E}_n \qquad \text{in vectorial form (2.53)} \tag{3.30}$$

Within each step, however, the Lagrangian stress remains related to a physically true Cauchy stress tensor by:

$$\boldsymbol{\sigma}_n = J_n \mathbf{F}_n \mathbf{S}_n \mathbf{F}_n^T \tag{3.31}$$

and using (2.24) the strain increment may be divided into linear and non-linear parts as (Lee 83):

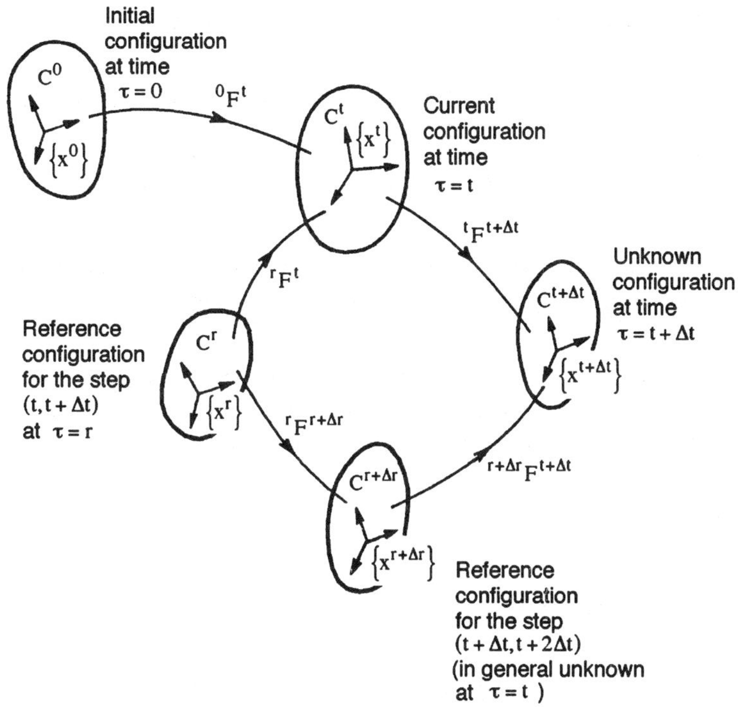

Fig. 3.4. Incremental description of the deformation process
(reprinted from (Kleiber 89) with his kind permission)

$$\Delta \mathbf{E} = \Delta \mathbf{E}^{L} + \Delta \mathbf{E}^{NL} \qquad \text{with} \tag{3.32}$$

$$\Delta \mathbf{E}^{NL} = \frac{1}{2}\left[\Delta\left(\nabla \mathbf{u}_n^T\right)\Delta\left(\nabla \mathbf{u}_n\right)\right] \qquad \text{as non-linear part} \tag{3.33}$$

$$\Delta \mathbf{E}^{L} = \frac{1}{2}\left[\Delta\left(\nabla \mathbf{u}_n\right) + \Delta\left(\nabla \mathbf{u}_n^T\right) + \Delta\left(\nabla \mathbf{u}_n^T\right)\nabla \mathbf{u}_n + \nabla \mathbf{u}_n^T \Delta\left(\nabla \mathbf{u}_n\right)\right] \text{ as linear part} \tag{3.34}$$

3.2.2 Incremental Formulation

Assuming all variables known at step n, the elementary Eulerian virtual work equation at step $n+1$ may be deduced from (2.48) in the form:

$$\delta \hat{W}_{n+1}^{acc} = \delta \hat{W}_{n+1}^{int} + \delta \hat{W}_{n+1}^{ext} \tag{3.35}$$

with $\delta\hat{W}^{acc}_{n+1} = \int_v \delta\hat{\mathbf{u}}^T_{n+1}\rho_{n+1}\mathbf{\Gamma}^v_{n+1}\,dv$ inertial virtual work

$\delta\hat{W}^{int}_{n+1} = -\int_v \mathrm{Tr}\left(\delta\hat{\mathbf{e}}^T_{n+1}\boldsymbol{\sigma}_{n+1}\right)dv = -\int_v \delta\hat{\boldsymbol{e}}^T_{n+1}\boldsymbol{\sigma}_{n+1}\,dv$ internal virtual work

$\delta\hat{W}^{ext}_{n+1} = \int_v \left(\delta\hat{\mathbf{u}}^f_{n+1}\right)^T\mathbf{f}^v_{n+1}\,dv + \int_s \left(\delta\hat{\mathbf{u}}^t_{n+1}\right)^T\mathbf{t}^s_{n+1}\mathbf{n}^s\,ds$ external virtual work

where $v \equiv v_{n+1}$, $s \equiv s_{n+1}$ and $\mathbf{n} = \mathbf{n}_{n+1}$ are the volume, surface and local normal vector to the external surface of the continuous medium at $n+1$.

Lagrangian Description. To formulate the incremental equation of motion, (3.35) must be moved into the previous configuration at n. The Lagrangian descriptions of the internal and inertial virtual works may be easily obtained as:

$$\delta\hat{W}^{acc}_{n+1} = \int_V \delta\hat{\mathbf{u}}^T_{n+1}\rho_n\mathbf{\Gamma}^V_{n+1}\,dV \tag{3.36}$$

$$\delta\hat{W}^{int}_{n+1} = -\int_V \delta\hat{\boldsymbol{E}}^T_{n+1}\boldsymbol{S}_{n+1}\,dV \tag{3.37}$$

where $V \equiv V_n$, $S \equiv S_n$ and $\mathbf{N} = \mathbf{N}_n$ are the volume, surface and local normal vector to the external surface of the continuous medium at n. By contrary, as the current configuration is unknown, no description of the external actions, applied on the current configuration at step $n+1$, can be defined in the previous configuration at step n. The assumption is then done that their descriptions in the configurations at step n and at step $n+1$ are close enough to take them as equal (Desai 80). Thus, the virtual work of external forces may be written as:

$$\delta\hat{W}^{ext}_{n+1} = \int_V \delta\hat{\mathbf{u}}^{fT}_{n+1}\mathbf{f}^V_{n+1}\,dV + \int_S \delta\hat{\mathbf{u}}^{tT}_{n+1}\mathbf{t}^S_{n+1}\mathbf{N}\,dS \tag{3.38}$$

Accounting for the fact that $\delta\mathbf{E}_{n+1} = \delta(\mathbf{E}_n + \Delta\mathbf{E}) = \delta(\Delta\mathbf{E})$, and using (3.29) and (3.32), the internal virtual work defined by (3.37) may be decomposed into:

$$\delta\hat{W}^{int}_{n+1} = \delta\hat{W}^{LS}_{n+1} + \delta\hat{W}^{L\Delta}_{n+1} + \delta\hat{W}^{NS}_{n+1} + \delta\hat{W}^{N\Delta}_{n+1} \quad \text{with} \tag{3.39}$$

$\delta\hat{W}^{LS}_{n+1} = -\int_V \delta\left(\Delta\hat{\boldsymbol{E}}^L\right)^T\boldsymbol{S}_n\,dV$ $\qquad$ $\delta\hat{W}^{L\Delta}_{n+1} = -\int_V \delta\left(\Delta\hat{\boldsymbol{E}}^L\right)^T\Delta\boldsymbol{S}\,dV$

$\delta\hat{W}^{NS}_{n+1} = -\int_V \delta\left(\Delta\hat{\boldsymbol{E}}^{NL}\right)^T\boldsymbol{S}_n\,dV$ $\qquad$ $\delta\hat{W}^{N\Delta}_{n+1} = -\int_V \delta\left(\Delta\hat{\boldsymbol{E}}^{NL}\right)^T\Delta\boldsymbol{S}\,dV$

Then, with a general incremental constitutive relationship in the form:

$$\Delta\mathbf{S} = \mathbf{K}_n^M \Delta\mathbf{E} + \mathbf{S}_n^R \qquad \text{or using (2.53):} \qquad \Delta\boldsymbol{S} = \boldsymbol{K}_n^M \Delta\boldsymbol{E} + \boldsymbol{S}_n^R \tag{3.40}$$

where $\mathbf{K}_n^M$ is the tangent material stiffness matrix depending on the current stress, current strain and past deformations

$\mathbf{S}_n^R$ is a residual constitutive stress term

(3.40), the virtual work component $\delta\hat{W}_{n+1}^{L\Delta}$ becomes:

$$\delta\hat{W}_{n+1}^{L\Delta} = -\int_V \delta\left(\Delta\hat{\boldsymbol{E}}^L\right)^T \left[\boldsymbol{K}_n^M \Delta\boldsymbol{E} + \boldsymbol{S}_n^R\right] dV \tag{3.41}$$

Finally, in order to linearize (3.39), the following assumptions are done (Lee 83):

$$\delta\hat{W}_{n+1}^{N\Delta} = -\int_V \delta\left(\Delta\hat{\boldsymbol{E}}^{NL}\right)^T \Delta\boldsymbol{S}\, dV \approx 0 \qquad \text{and} \tag{3.42}$$

$$\Delta\mathbf{S} \approx \Delta\mathbf{S}\left(\Delta\mathbf{E}^L\right) \qquad \text{i.e.,} \qquad \delta\hat{W}_{n+1}^{L\Delta} \approx -\int_V \delta\left(\Delta\hat{\boldsymbol{E}}^L\right)^T \left[\boldsymbol{K}_n^M \Delta\boldsymbol{E}^L + \boldsymbol{S}_n^R\left(\Delta\boldsymbol{E}^L\right)\right] dV$$

As a result, (3.35) may be written in the form:

$$\delta\hat{W}_{n+1}^{acc} = \delta\hat{W}_{n+1}^1 + \delta\hat{W}_{n+1}^2 + \delta\hat{W}_{n+1}^3 + \delta\hat{W}_{n+1}^{ext} \tag{3.43}$$

with:

$$\delta\hat{W}_{n+1}^{acc} = \int_V \delta\hat{\mathbf{u}}_{n+1}^T \rho_n \boldsymbol{\Gamma}_{n+1}^V\, dV \qquad \text{inertial virtual work}$$

$$\delta\hat{W}_{n+1}^1 = -\int_V \delta\left(\Delta\hat{\boldsymbol{E}}^L\right)^T \left[\boldsymbol{S}_n + \boldsymbol{S}_n^R\right] dV \qquad \text{internal force virtual work}$$

$$\delta\hat{W}_{n+1}^2 = -\int_V \delta\left(\Delta\hat{\boldsymbol{E}}^L\right)^T \boldsymbol{K}_n^M \Delta\boldsymbol{E}^L\, dV \qquad \text{linear strain virtual work}$$

$$\delta\hat{W}_{n+1}^3 = -\int_V \delta\left(\Delta\hat{\boldsymbol{E}}^{NL}\right)^T \left[\boldsymbol{S}_n + \boldsymbol{S}_n^R\right] dV \qquad \text{geometric virtual work}$$

$$\delta\hat{W}_{n+1}^{ext} = \int_V \delta\hat{\mathbf{u}}_{n+1}^{fT} \mathbf{f}_{n+1}^V\, dV + \int_S \delta\hat{\mathbf{u}}_{n+1}^{tT} \mathbf{t}_{n+1}^S \mathbf{N}\, dS \qquad \text{external virtual work}$$

3.2.3 Finite Element Formulation

Assuming a finite element discretization as defined in (3.1), i.e.:

$$\mathbf{u}_n = \mathbf{H}\,\mathbf{U}_n \qquad \mathbf{u}^f_n = \mathbf{H}^f\mathbf{U}_n \qquad \mathbf{u}^t_n = \mathbf{H}^t\mathbf{U}_n \tag{3.44}$$

the strain-displacement relations may be written for both strain components as:

$$\Delta\mathbf{E}^L = \mathbf{B}^L_n\Delta\mathbf{U} \qquad \boldsymbol{\psi}\,\Delta\mathbf{E}^{NL} = \Delta\mathbf{U}^T\,\mathbf{B}^{NL^T}_n\boldsymbol{\Psi}\mathbf{B}^{NL}_n\,\Delta\mathbf{U} \quad \text{with} \tag{3.45}$$

$$\boldsymbol{\Psi} = \begin{bmatrix} \boldsymbol{\psi} & [0] & [0] \\ [0] & \ddots & [0] \\ [0] & [0] & \boldsymbol{\psi} \end{bmatrix} \qquad \text{for any matrix } \boldsymbol{\psi} \text{ (Lee 83),} \qquad \text{i.e., using (2.53)}$$

$$\Delta\boldsymbol{E}^L = \boldsymbol{B}^L_n\Delta\mathbf{U} \qquad \boldsymbol{\psi}\,\Delta\boldsymbol{E}^{NL} = \Delta\mathbf{U}^T\,\boldsymbol{B}^{NL^T}_n\boldsymbol{\Psi}\boldsymbol{B}^{NL}_n\,\Delta\mathbf{U} \tag{3.46}$$

where $\Delta\mathbf{U}$ is the nodal displacement increment at step n. Then, replacing (3.44), (3.45), and (3.46) in (3.43), the virtual work equation may be converted into:

$$\delta\left(\Delta\hat{\mathbf{U}}\right)^T\left[\mathbf{M}_n\ddot{\mathbf{U}}_{n+1} + \mathbf{K}^L_n\Delta\mathbf{U} + \mathbf{K}^{NL}_n\Delta\hat{\mathbf{U}}\right] = \delta\left(\Delta\hat{\mathbf{U}}\right)^T\left[\mathbf{L}_{n+1} - \mathbf{R}_n\right] \qquad \text{i.e.:}$$

$$\mathbf{M}_n\ddot{\mathbf{U}}_{n+1} + \left(\mathbf{K}^L_n + \mathbf{K}^{NL}_n\right)\Delta\mathbf{U} = \mathbf{L}_{n+1} - \mathbf{R}_n \qquad \text{for real work } \left(\Delta\hat{\mathbf{U}} \equiv \Delta\mathbf{U}\right) \tag{3.47}$$

with:

$$\mathbf{M}_n = \int_V \rho_n\mathbf{H}^T\mathbf{H}\,dV \qquad \text{as tangent mass matrix}$$

$$\mathbf{K}^L_n = \int_V \boldsymbol{B}^{L^T}_n\boldsymbol{K}^M_n\boldsymbol{B}^L_n\,dV \qquad \text{as linear tangent stiffness matrix}$$

$$\mathbf{K}^{NL}_n = \int_V \boldsymbol{B}^{NL^T}_n\left[\boldsymbol{S}_n + \boldsymbol{S}^R_n\right]\boldsymbol{B}^{NL}_n\,dV \qquad \text{as the geometric stiffness matrix}$$

$$\mathbf{R}_n = \int_V \boldsymbol{B}^{L^T}_n\left[\boldsymbol{S}_n + \boldsymbol{S}^R_n\right]dV \qquad \text{as internal force vector}$$

$$\mathbf{L}_{n+1} = \int_V \mathbf{H}^{f^T}\mathbf{f}^V_{n+1}\,dV + \int_S \mathbf{H}^{t^T}\mathbf{t}^S_{n+1}\mathbf{N}\,dS \quad \text{as the external force vector}$$

Solution may be simplified by neglecting inertia effects $\mathbf{M}_n\ddot{\mathbf{U}}_{n+1} \approx [0]$ (Lee 83).

3.2.4 Small Deformations

When strains are small, Eulerian and Lagrangian configurations are close, so that the Cauchy stress tensor $\boldsymbol{\sigma}$ and the infinitesimal strain tensor $\boldsymbol{\epsilon}$ may be considered as state tensors within the same linear equation of motion, instead of the Kirchhoff second stress $\mathbf{S}$ and Green–Lagrange strain $\mathbf{E}$ tensors as previously. But in case of a non-linear material, an incremental formulation is still required to achieve the analysis (Katona 78).

Incremental Motion. As the strain is small, the general virtual work equation defined by (3.43) becomes:

$$\delta\hat{W}^{acc}_{n+1} = \delta\hat{W}^{1}_{n+1} + \delta\hat{W}^{2}_{n+1} + \delta\hat{W}^{ext}_{n+1} \tag{3.48}$$

with

$$\delta\hat{W}^{acc}_{n+1} = \int_V \delta\hat{\mathbf{u}}^T_{n+1}\rho_n \boldsymbol{\Gamma}^V_{n+1}\, dV \quad \text{inertial virtual work}$$

$$\delta\hat{W}^{1}_{n+1} = -\int_V \delta(\Delta\hat{\boldsymbol{\epsilon}})^T\left[\boldsymbol{\sigma}_n + \boldsymbol{\sigma}^R_n\right] dV \quad \text{internal force virtual work}$$

$$\delta\hat{W}^{2}_{n+1} = -\int_V \delta(\Delta\hat{\boldsymbol{\epsilon}})^T \boldsymbol{K}^M_n \Delta\boldsymbol{\epsilon}\, dV \quad \text{small strain virtual work}$$

$$\delta\hat{W}^{ext}_{n+1} = \int_V \delta\hat{\mathbf{u}}^{t^T}_{n+1}\mathbf{f}^V_{n+1}\, dV + \int_S \delta\hat{\mathbf{u}}^{t^T}_{n+1}\mathbf{t}^S_{n+1}\mathbf{N}\, dS \quad \text{external virtual work}$$

Finite Element Formulation. As there are no geometrical non-linearities, the finite element equation dynamics defined in (3.47) becomes:

$$\mathbf{M}\,\ddot{\mathbf{U}}_{n+1} + \mathbf{K}_n\Delta\mathbf{U} = \mathbf{L}_{n+1} - \mathbf{R}_n \tag{3.49}$$

with

$$\mathbf{M} = \int_V \rho_0 \mathbf{H}^T\mathbf{H}\, dV \quad \text{as tangent mass matrix}$$

$$\mathbf{K}_n = \int_V \boldsymbol{B}^T\boldsymbol{K}^M_n\boldsymbol{B}\, dV \quad \text{as tangent stiffness matrix}$$

$$\mathbf{R}_n = \int_V \boldsymbol{B}^T\left[\boldsymbol{\sigma}_n + \boldsymbol{\sigma}^R_n\right] dV \quad \text{as internal force vector}$$

$$\mathbf{L}_{n+1} = \int_V \mathbf{H}^{t^T}\mathbf{f}^V_{n+1}\, dV + \int_S \mathbf{H}^{t^T}\mathbf{t}^S_{n+1}\mathbf{N}\, dS \quad \text{as the external force vector}$$

Solution may be simplified by neglecting inertia: $\mathbf{M}_n\ddot{\mathbf{U}}_{n+1} \approx [0]$ (Kleiber 89).

3.2.5 Linear Viscoelasticity

As noticed by Katona, the series form of linear viscoelastic defined in (2.117) for infinitesimal strains, is equally valid for large deformations provided they are used in a consistent fashion with Lagrangian variables (Katona 78). Thus, in the following, we propose the development of the incremental viscoelasticity for finite deformation, based on Katona's development for infinitesimal strains presented in (Katona 80).

Incremental Law. From (2.119), an incremental linear viscoelastic constitutive relation may be derived in the form:

$$\Delta\mathbf{S} = \mathbf{K}^{E}\Delta\mathbf{E} - \mathbf{D}^{P}\sum_{i=1}^{n_P} P_i\,\Delta\mathbf{P}_i - \mathbf{D}^{Q}\sum_{i=1}^{n_Q} Q_i\,\Delta\mathbf{Q}_i \qquad \text{with} \tag{3.50}$$

$$\Delta\mathbf{P}_i = \int_{t_n}^{t_{n+1}} \frac{\mathbf{E}(\tau)}{\beta_i} e^{-\frac{t_{n+1}-\tau}{\beta_i}} d\tau - \left(1 - e^{-\frac{\Delta t}{\beta_i}}\right)\mathbf{P}_{i\_n} \qquad \Delta\mathbf{P}_i = \mathbf{P}_{i\_n+1} - \mathbf{P}_{i\_n} \tag{3.51}$$

$$\Delta\mathbf{Q}_i = \int_{t_n}^{t_{n+1}} \frac{\mathbf{E}(\tau)}{\gamma_i} e^{-\frac{t_{n+1}-\tau}{\gamma_i}} d\tau - \left(1 - e^{-\frac{\Delta t}{\gamma_i}}\right)\mathbf{Q}_{i\_n} \qquad \Delta\mathbf{Q}_i = \mathbf{Q}_{i\_n+1} - \mathbf{Q}_{i\_n} \tag{3.52}$$

With a linear interpolation assumption for strain over the time step interval:

$$\mathbf{E}(\tau) = \mathbf{E}_n + \frac{\tau - t_n}{\Delta t}\Delta\mathbf{E} \tag{3.53}$$

integrations of (3.51) and (3.52) result in:

$$\Delta\mathbf{P}_i = \left[1 - \frac{\beta_i}{\Delta t}\left(1 - e^{-\frac{\Delta t}{\beta_i}}\right)\right]\Delta\mathbf{E} + \left(1 - e^{-\frac{\Delta t}{\beta_i}}\right)\left[\mathbf{E}_n - \mathbf{P}_{i\_n}\right] \tag{3.54}$$

$$\Delta\mathbf{Q}_i = \left[1 - \frac{\gamma_i}{\Delta t}\left(1 - e^{-\frac{\Delta t}{\gamma_i}}\right)\right]\Delta\mathbf{E} + \left(1 - e^{-\frac{\Delta t}{\gamma_i}}\right)\left[\mathbf{E}_n - \mathbf{Q}_{i\_n}\right] \tag{3.55}$$

Replacing (3.54) and (3.55) into (3.50) leads to an incremental relation as:

$$\Delta\mathbf{S} = \left[\mathbf{K}^{E} - \mathbf{K}^{V}\right]\Delta\mathbf{E} - \mathbf{S}_n^{V} \quad \text{or using (2.53):} \quad \Delta\boldsymbol{S} = \left[\boldsymbol{K}^{E} - \boldsymbol{K}^{V}\right]\Delta\boldsymbol{E} - \boldsymbol{S}_n^{V} \tag{3.56}$$

with $\mathbf{K}^{E} = \mathbf{\Phi}(0) = \mathbf{D}^{P} P(0) + \mathbf{D}^{Q} Q(0)$

$$\mathbf{K}^{V} = \mathbf{D}^{P} \sum_{i=1}^{n_P} P_i \left[1 - \frac{\beta_i}{\Delta t}\left(1 - e^{-\frac{\Delta t}{\beta_i}}\right)\right] + \mathbf{D}^{Q} \sum_{i=1}^{n_Q} Q_i \left[1 - \frac{\gamma_i}{\Delta t}\left(1 - e^{-\frac{\Delta t}{\gamma_i}}\right)\right]$$

$$\mathbf{S}_n^{V} = \mathbf{D}^{P} \sum_{i=1}^{n_P} P_i \left(1 - e^{-\frac{\Delta t}{\beta_i}}\right)\left[\mathbf{E}_n - \mathbf{P}_{i\_n}\right] + \mathbf{D}^{Q} \sum_{i=1}^{n_Q} Q_i \left(1 - e^{-\frac{\Delta t}{\gamma_i}}\right)\left[\mathbf{E}_n - \mathbf{Q}_{i\_n}\right]$$

$\mathbf{K}^{E}$ and $\mathbf{K}^{V}$ are the elastic and viscous matrix, $\mathbf{S}_n^{V}$ is the viscous stress-history influence vector. The viscous matrix $\mathbf{K}^{V}$ only depends on the time step size. The viscous stress-history influence vector $\mathbf{S}_n^{V}$ accounts for the influence of all past deformations on the current stress increment (Katona 80). By referring to (3.40):

$$\mathbf{K}_n^{M} = \mathbf{K}^{E} - \mathbf{K}^{V} \qquad \text{and} \qquad \mathbf{S}_n^{R} = \mathbf{S}_n^{V} \tag{3.57}$$

Incremental Motion. With (3.56), the virtual work equation (3.43) becomes:

$$\delta \hat{W}_{n+1}^{acc} = \delta \hat{W}_{n+1}^{1} + \delta \hat{W}_{n+1}^{2} + \delta \hat{W}_{n+1}^{3} + \delta \hat{W}_{n+1}^{ext} \tag{3.58}$$

with $\delta \hat{W}_{n+1}^{acc} = \int_V \delta \hat{\mathbf{u}}_{n+1}^{T} \rho_n \mathbf{\Gamma}_{n+1}^{V}\, dV$ inertial virtual work

$$\delta \hat{W}_{n+1}^{1} = -\int_V \delta\left(\Delta \hat{\boldsymbol{E}}^{L}\right)^{T}\left[\boldsymbol{S}_n + \boldsymbol{S}_n^{R}\right] dV \qquad \text{internal force virtual work}$$

$$\delta \hat{W}_{n+1}^{2} = -\int_V \delta\left(\Delta \hat{\boldsymbol{E}}^{L}\right)^{T}\left[\boldsymbol{K}^{E} - \boldsymbol{K}^{V}\right] \Delta \boldsymbol{E}^{L}\, dV \qquad \text{linear strain virtual work}$$

$$\delta \hat{W}_{n+1}^{3} = -\int_V \delta\left(\Delta \hat{\boldsymbol{E}}^{NL}\right)^{T}\left[\boldsymbol{S}_n + \boldsymbol{S}_n^{V}\right] dV \qquad \text{geometric virtual work}$$

$$\delta \hat{W}_{n+1}^{ext} = \int_V \delta \hat{\mathbf{u}}_{n+1}^{fT} \mathbf{f}_{n+1}^{V}\, dV + \int_S \delta \hat{\mathbf{u}}_{n+1}^{tT} \mathbf{t}_{n+1}^{S} \mathbf{N}\, dS \qquad \text{external virtual work}$$

Finite Element Formulation. As defined in (3.47), the finite element equation dynamics is obtained in the form:

$$\mathbf{M}_n \ddot{\mathbf{U}}_{n+1} + \left(\mathbf{K}_n^{L} + \mathbf{K}_n^{NL}\right) \Delta \mathbf{U} = \mathbf{L}_{n+1} - \mathbf{R}_n \qquad \text{for real work } \left(\Delta \hat{\mathbf{U}} \equiv \Delta \mathbf{U}\right) \tag{3.59}$$

with $\mathbf{M}_n = \int_V \rho_n \mathbf{H}^T\mathbf{H}\, dV$ as tangent mass matrix

$$\mathbf{K}_n^L = \int_V \boldsymbol{B}_n^{L^T} \left[\boldsymbol{K}^E - \boldsymbol{K}^V\right] \boldsymbol{B}_n^L\, dV$$ as linear tangent stiffness matrix

$$\mathbf{K}_n^{NL} = \int_V \boldsymbol{B}_n^{NL^T} \left[\boldsymbol{S}_n + \boldsymbol{S}_n^V\right] \boldsymbol{B}_n^{NL}\, dV$$ as geometric stiffness matrix

$$\mathbf{R}_n = \int_V \boldsymbol{B}_n^{L^T} \left[\boldsymbol{S}_n + \boldsymbol{S}_n^V\right] dV$$ as internal force vector

$$\mathbf{L}_{n+1} = \int_V \mathbf{H}^{f^T} \mathbf{f}_{n+1}^V\, dV + \int_S \mathbf{H}^{t^T} \mathbf{t}_{n+1}^S \mathbf{N}\, dS$$ as external force vector

Solution may be simplified by neglecting inertia effects: $\mathbf{M}_n \ddot{\mathbf{U}}_{n+1} \approx [0]$ (Lee 83).

Force Vector Decomposition. In order to improve the computational process, it is convenient to decompose the internal force vector $\mathbf{R}_n$ into:

$$\mathbf{R}_n = \mathbf{R}_n^S + \mathbf{R}_n^V \quad \text{with} \tag{3.60}$$

$$\mathbf{R}_n^S = \int_V \boldsymbol{B}_n^{L^T} \boldsymbol{S}_n\, dV \quad \text{and} \quad \mathbf{R}_n^V = \int_V \boldsymbol{B}_n^{L^T} \boldsymbol{S}_n^V\, dV$$

and to formulate the viscoelastic force history increment vector $\Delta\mathbf{R}_n^V$ as a function of the displacement increment $\Delta\mathbf{U}$. Using $\boldsymbol{S}_n^V$, vectorial form of $\mathbf{S}_n^V$ defined in (3.56), the viscoelastic force history vector $\mathbf{R}_n^V$ is decomposed into (Katona 78):

$$\mathbf{R}_n^V = \sum_{i=1}^{n_P} \mathbf{h}_n^{P_i} + \sum_{i=1}^{n_Q} \mathbf{h}_n^{Q_i} \quad \text{with} \tag{3.61}$$

$$\mathbf{h}_n^{P_i} = P_i \left(1 - e^{-\frac{\Delta t}{\beta_i}}\right) \int_V \boldsymbol{B}_n^{L^T} \boldsymbol{D}^P \left[\boldsymbol{E}_n - \boldsymbol{P}_{i\_n}\right] dV \tag{3.62}$$

$$\mathbf{h}_n^{Q_i} = Q_i \left(1 - e^{-\frac{\Delta t}{\gamma_i}}\right) \int_V \boldsymbol{B}_n^{L^T} \boldsymbol{D}^Q \left[\boldsymbol{E}_n - \boldsymbol{Q}_{i\_n}\right] dV \tag{3.63}$$

From (3.62) and (3.63), the increments $\Delta\mathbf{h}_n^{P_i}$ and $\Delta\mathbf{h}_n^{Q_i}$ may be derived as:

$$\Delta \mathbf{h}^{P_i} = P_i \left(1 - e^{-\frac{\Delta t}{\beta_i}}\right) \int_V \boldsymbol{B}_n^{L^T} \boldsymbol{D}^P \left[\Delta \boldsymbol{E} - \Delta \boldsymbol{P}_i\right] dV \tag{3.64}$$

$$\Delta \mathbf{h}^{Q_i} = Q_i \left(1 - e^{-\frac{\Delta t}{\gamma_i}}\right) \int_V \boldsymbol{B}_n^{L^T} \boldsymbol{D}^Q \left[\Delta \boldsymbol{E} - \Delta \boldsymbol{Q}_i\right] dV \tag{3.65}$$

Then, using (3.42), (3.51) and (3.52), $\Delta \mathbf{h}_n^{P_i}$ and $\Delta \mathbf{h}_n^{Q_i}$ may be developed into:

$$\Delta \mathbf{h}^{P_i} = P_i \frac{\beta_i}{\Delta t} \left(1 - e^{-\frac{\Delta t}{\beta_i}}\right)^2 \mathbf{K}^P \Delta \mathbf{U} - \left(1 - e^{-\frac{\Delta t}{\beta_i}}\right) \mathbf{h}_n^{P_i} \qquad \mathbf{K}^P = \int_V \boldsymbol{B}_n^{L^T} \boldsymbol{D}^P \boldsymbol{B}_n^L \, dV \tag{3.66}$$

$$\Delta \mathbf{h}^{Q_i} = Q_i \frac{\gamma_i}{\Delta t} \left(1 - e^{-\frac{\Delta t}{\gamma_i}}\right)^2 \mathbf{K}^Q \Delta \mathbf{U} - \left(1 - e^{-\frac{\Delta t}{\gamma_i}}\right) \mathbf{h}^{Q_i} \qquad \mathbf{K}^Q = \int_V \boldsymbol{B}_n^{L^T} \boldsymbol{D}^Q \boldsymbol{B}_n^L \, dV \tag{3.67}$$

with $\mathbf{K}^P$ and $\mathbf{K}^Q$ as dimensionless *bulk* and *shear stiffness* matrices. As a result, the viscoelastic force history vector $\mathbf{R}_n^V$ may be updated at each step, using:

$$\mathbf{R}_{n+1}^V = \mathbf{R}_n^V + \Delta \mathbf{R}^V \qquad \text{with} \qquad \Delta \mathbf{R}^V = \sum_{i=1}^{n_P} \Delta \mathbf{h}^{P_i} + \sum_{i=1}^{n_Q} \Delta \mathbf{h}^{Q_i} \tag{3.68}$$

by just computing the increment $\Delta \mathbf{R}^V$ for the nodal displacement increment $\Delta \mathbf{U}$. using (3.66) and (3.67) (Katona 80).

Small Strains. As the strain is small, the virtual work equation defined by (3.58) becomes:

$$\delta \hat{W}_{n+1}^{acc} = \delta \hat{W}_{n+1}^1 + \delta \hat{W}_{n+1}^2 + \delta \hat{W}_{n+1}^{ext} \tag{3.69}$$

with

$$\delta \hat{W}_{n+1}^{acc} = \int_V \delta \hat{\mathbf{u}}_{n+1}^T \rho_n \boldsymbol{\Gamma}_{n+1}^V \, dV \qquad \text{inertial virtual work}$$

$$\delta \hat{W}_{n+1}^1 = -\int_V \delta(\Delta \hat{\boldsymbol{\epsilon}})^T \left[\boldsymbol{\sigma}_n + \boldsymbol{\sigma}_n^V\right] dV \qquad \text{internal force virtual work}$$

$$\delta \hat{W}_{n+1}^2 = -\int_V \delta(\Delta \hat{\boldsymbol{\epsilon}})^T \left[\boldsymbol{K}^E - \boldsymbol{K}^V\right] \Delta \boldsymbol{\epsilon} \, dV \qquad \text{small strain virtual work}$$

$$\delta \hat{W}_{n+1}^{ext} = \int_V \delta \hat{\mathbf{u}}_{n+1}^{f^T} \mathbf{f}_{n+1}^V \, dV + \int_S \delta \hat{\mathbf{u}}_{n+1}^{t^T} \mathbf{t}_{n+1}^S \mathbf{N} \, dS \qquad \text{external virtual work}$$

and the finite element equation dynamics defined in (3.59) becomes:

$$\mathbf{M}\,\ddot{\mathbf{U}}_{n+1} + \mathbf{K}\,\Delta\mathbf{U} = \mathbf{L}_{n+1} - \mathbf{R}_n \tag{3.70}$$

with $\mathbf{M} = \int_V \rho_0 \mathbf{H}^T\mathbf{H}\,dV$ as tangent mass matrix

$\mathbf{K} = \int_V \boldsymbol{B}^T\left[\boldsymbol{K}^E - \boldsymbol{K}^V\right]\boldsymbol{B}\,dV$ as tangent stiffness matrix

$\mathbf{R}_n = \int_V \boldsymbol{B}^T\left[\boldsymbol{\sigma}_n + \boldsymbol{\sigma}_n^V\right]dV$ as internal force vector

$\mathbf{L}_{n+1} = \int_V \mathbf{H}^{fT}\mathbf{f}_{n+1}^V\,dV + \int_S \mathbf{H}^{tT}\mathbf{t}_{n+1}^S\mathbf{N}\,dS$ as the external force vector

and with a viscoelastic force-history increment vector derived from (3.68) as:

$$\mathbf{R}_{n+1}^V = \mathbf{R}_n^V + \Delta\mathbf{R}^V \quad \text{with} \quad \Delta\mathbf{R}^V = \sum_{i=1}^{n_P}\Delta\mathbf{h}^{P_i} + \sum_{i=1}^{n_Q}\Delta\mathbf{h}^{Q_i} \tag{3.71}$$

$$\Delta\mathbf{h}^{P_i} = P_i\,\frac{\beta_i}{\Delta t}\left(1-e^{-\frac{\Delta t}{\beta_i}}\right)^2\mathbf{K}^P\Delta\mathbf{U} - \left(1-e^{-\frac{\Delta t}{\beta_i}}\right)\mathbf{h}_n^{P_i} \qquad \mathbf{K}^P = \int_V \boldsymbol{B}^T\boldsymbol{D}^P\boldsymbol{B}\,dV \tag{3.72}$$

$$\Delta\mathbf{h}^{Q_i} = Q_i\,\frac{\gamma_i}{\Delta t}\left(1-e^{-\frac{\Delta t}{\gamma_i}}\right)^2\mathbf{K}^Q\Delta\mathbf{U} - \left(1-e^{-\frac{\Delta t}{\gamma_i}}\right)\mathbf{h}_n^{Q_i} \qquad \mathbf{K}^Q = \int_V \boldsymbol{B}^T\boldsymbol{D}^Q\boldsymbol{B}\,dV \tag{3.73}$$

Solution may be simplified by neglecting inertia: $\mathbf{M}_n\ddot{\mathbf{U}}_{n+1} \approx [0]$ (Katona 78).

3.3 Incremental Resolution

3.3.1 Incremental Formulation

Considering non-linear dynamics problems, the general incremental finite element equation of motion may be carried out from (3.14) as follows (Curnier 94):

$$d\left[\mathbf{M}\ddot{\mathbf{U}} + \boldsymbol{\Pi}\left(\mathbf{U},\dot{\mathbf{U}}\right) - \mathbf{L}\right] = \mathbf{M}\,d\ddot{\mathbf{U}} + \mathbf{D}\,d\dot{\mathbf{U}} + \mathbf{K}\,d\mathbf{U} - d\mathbf{L} \quad \text{i.e.:} \tag{3.74}$$

$$\Delta\left[\mathbf{M}\ddot{\mathbf{U}} + \boldsymbol{\Pi}\left(\mathbf{U},\dot{\mathbf{U}}\right) - \mathbf{L}\right] = \mathbf{M}_n\Delta\ddot{\mathbf{U}} + \mathbf{D}_n\Delta\dot{\mathbf{U}} + \mathbf{K}_n\Delta\mathbf{U} - \Delta\mathbf{L} \quad \text{i.e.:} \tag{3.75}$$

$$\mathbf{M}_n \Delta\ddot{\mathbf{U}} + \mathbf{D}_n \Delta\dot{\mathbf{U}} + \mathbf{K}_n \Delta\mathbf{U} = \Delta\mathbf{L} \tag{3.76}$$

with

	$\mathbf{M}_n$	tangent mass matrix
	$\mathbf{D}_n$	tangent damping matrix
	$\mathbf{K}_n$	tangent stiffness matrix
	$\Delta\mathbf{L}$	nodal external force increment vector for step n
	$\Delta\mathbf{U}$	nodal displacement increment vector for step n

Taking into account the incremental relations of the nodal displacement vector as:

$$\Delta\mathbf{U} = \mathbf{U}_{n+1} - \mathbf{U}_n \qquad \Delta\dot{\mathbf{U}} = \dot{\mathbf{U}}_{n+1} - \dot{\mathbf{U}}_n \qquad \Delta\ddot{\mathbf{U}} = \ddot{\mathbf{U}}_{n+1} - \ddot{\mathbf{U}}_n \tag{3.77}$$

(3.76) may be developed into:

$$\mathbf{M}_n \ddot{\mathbf{U}}_{n+1} + \mathbf{D}_n \dot{\mathbf{U}}_{n+1} + \mathbf{K}_n \Delta\mathbf{U} = \mathbf{L}_{n+1} - \mathbf{R}_n \qquad \text{with} \tag{3.78}$$

$$\mathbf{R}_n = \mathbf{L}_n - \mathbf{M}_n \ddot{\mathbf{U}}_n - \mathbf{D}_n \dot{\mathbf{U}}_n \qquad \text{and} \qquad \mathbf{R}_{n+1} = \mathbf{R}_n + \mathbf{K}_n \Delta\mathbf{U} \tag{3.79}$$

which may be used as a basis for the development of incremental algorithms (Kleiber 89). All components are assumed known up to step n, except the nodal displacement increment vector $\Delta\mathbf{U}$ which is sought for updating up to step n+1.

3.3.2 The Finite Difference Methods

The finite difference methods are based on finite difference approximations of the relationships between the acceleration, the velocity, and the unknown incremental displacement vector $\mathbf{U}_n$. There are mainly two groups of integration approaches.

Explicit Integration. The acceleration and velocity vectors may be expressed as:

$$\ddot{\mathbf{U}}_n = \frac{1}{\Delta t^2}\left(\mathbf{U}_{n+1} - 2\mathbf{U}_n + \mathbf{U}_{n-1}\right) = \frac{1}{\Delta t^2}\left[\Delta\mathbf{U} - \left(\mathbf{U}_n - \mathbf{U}_{n-1}\right)\right] \tag{3.80}$$

$$\dot{\mathbf{U}}_n = \frac{1}{2\Delta t}\left(\mathbf{U}_{n+1} - \mathbf{U}_{n-1}\right) = \frac{1}{2\Delta t}\left[\Delta\mathbf{U} + \left(\mathbf{U}_n - \mathbf{U}_{n-1}\right)\right] \tag{3.81}$$

Substituting (3.80) and (3.81) into the equation (3.78) written for $n-1$, the incremental dynamics may be obtained in the form:

$$\mathbf{A}_n \Delta \mathbf{U} = \mathbf{L}_n - \mathbf{Y}_n \qquad \text{with} \tag{3.82}$$

$$\mathbf{A}_n = \frac{1}{\Delta t^2}\mathbf{M}_n + \frac{1}{2\Delta t}\mathbf{D}_n \qquad \mathbf{Y}_n = \mathbf{R}_n - \left(\frac{1}{\Delta t^2}\mathbf{M}_n - \frac{1}{2\Delta t}\mathbf{D}_n\right)\left(\mathbf{U}_n - \mathbf{U}_{n-1}\right)$$

where: $\mathbf{A}_n$ is the effective stiffness matrix at step n
$\mathbf{L}_n$ is the nodal external force vector at step n
$\mathbf{Y}_n$ is the effective internal force vector at step n

As step n is known, (3.82) may be solved for the nodal increment vector $\Delta\mathbf{U}$, and all the tangent matrix may be updated before considering the next step. In case the mass and damping matrix are diagonal, such an integration scheme does not require a factorization of the effective stiffness matrix $\mathbf{A}_n$. Since no matrices of the complete structure have to be calculated, the solution can essentially be carried out at the element level and relatively little high-speed storage is required. This explicit method is usually known as the *central difference* method. It is said to be conditionally stable, i.e. the time step must be smaller than the critical value which may be calculated from the mass, damping and stiffness properties of the complete element assemblage. A noteworthy point is that the step-by-step solution scheme based on the explicit approach does not reduce to static analysis, if the inertial and viscous effects are neglected (Kleiber 89).

Implicit Integration. Using two parameters α and β, the nodal displacement and velocity vectors at step $n+1$ may be expressed as follows:

$$\mathbf{U}_{n+1} = \mathbf{U}_n + \dot{\mathbf{U}}_n\,\Delta t + \left[\left(\frac{1}{2} - \alpha\right)\ddot{\mathbf{U}}_n + \alpha\,\ddot{\mathbf{U}}_{n+1}\right]\Delta t^2 \tag{3.83}$$

$$\dot{\mathbf{U}}_{n+1} = \dot{\mathbf{U}}_n + \Delta t\left[(1-\beta)\,\ddot{\mathbf{U}}_n + \beta\,\ddot{\mathbf{U}}_{n+1}\right] \tag{3.84}$$

Extracting $\ddot{\mathbf{U}}_{n+1}$, $\dot{\mathbf{U}}_{n+1}$ from (3.83), (3.84), and substituting them in (3.78) gives:

$$\mathbf{A}_n \Delta \mathbf{U} = \mathbf{L}_{n+1} - \mathbf{Y}_n \qquad \text{with} \tag{3.85}$$

$$\mathbf{A}_n = \frac{1}{\alpha\Delta t^2}\mathbf{M}_n + \frac{\beta}{\alpha\Delta t}\mathbf{D}_n + \mathbf{K}_n$$

$$\mathbf{Y}_n = \mathbf{R}_n - \mathbf{M}_n\left[\frac{1}{\alpha\Delta t}\dot{\mathbf{U}}_n + \left(\frac{1}{2\alpha} - 1\right)\ddot{\mathbf{U}}_n\right] - \mathbf{D}_n\left[\left(\frac{\beta}{\alpha} - 1\right)\dot{\mathbf{U}}_n + \frac{\Delta t}{2}\left(\frac{\beta}{\alpha} - 2\right)\ddot{\mathbf{U}}_n\right]$$

where: $\mathbf{A}_n$ is the effective stiffness matrix at step n
$\mathbf{L}_{n+1}$ is the nodal external force vector at step n+1
$\mathbf{Y}_n$ is the effective internal force vector at step n
α and β are parameters for tuning stability and accuracy

A basic difference between this approach and the previous one is that, since the equilibrium is considered at time $t+\Delta t$, the tangent stiffness matrix $\mathbf{K}_n$ appears as factor to the required incremental displacement $\Delta \mathbf{U}$. This method is known as the *Newmark method.* Another difference is also that this can be used for static problems in neglecting the inertial and viscous effects. Furthermore, for some values of parameters α and β, there is no critical time-step limit, and a much larger time-step Δt can be used (Kleiber 89).

3.3.3 The Linear Iteration Methods

In the explicit approaches such as the central difference method presented in (3.82), the solution of the incremental finite element equations may be considered entirely consistent with the dynamic equilibrium conditions at time t, since the configuration is known at this time. Therefore, no further improvement is necessary. In the implicit approaches, on the other hand, the configuration at time $t+\Delta t$, for which the equilibrium conditions are established, is unknown. This makes it usually necessary to perform additional iterations, in order to find a more accurate solution *within each time interval.* This may be achieved by means of a *Newton–Raphson* iteration scheme (Kleiber 89).

Newton-Raphson Scheme. This scheme consists in seeking the solution of:

$$\mathbf{M}_n \ddot{\mathbf{U}}_{n+1\_i} + \mathbf{D}_n \dot{\mathbf{U}}_{n+1\_i} + \mathbf{K}_{n+1\_i} \Delta \mathbf{U}_i = \mathbf{L}_{n+1} - \mathbf{R}_{n+1\_i-1} \qquad i \equiv (1,2,3,...) \qquad \text{with} \quad (3.86)$$

$$\mathbf{K}_{n+1\_0} = \mathbf{K}_n \qquad \mathbf{R}_{n+1\_0} = \mathbf{R}_n \qquad \mathbf{U}_{n+1\_0} = \mathbf{U}_n \qquad \text{as initial conditions}$$

in which the nodal velocity and acceleration vectors $\dot{\mathbf{U}}_{n+1}$ and $\ddot{\mathbf{U}}_{n+1}$ are defined by (3.80) and (3.81), and the internal nodal force vector $\mathbf{R}_{n+1\_i-1}$ corresponds to the state of stress related to the configuration with the displacements as:

$$\mathbf{U}_{n+1\_i-1} = \mathbf{U}_{n+1\_i-2} + \Delta \mathbf{U}_{i-1} \qquad (3.87)$$

where $\Delta \mathbf{U}_{i-1}$ is the i-th correction to the incremental displacement vector.

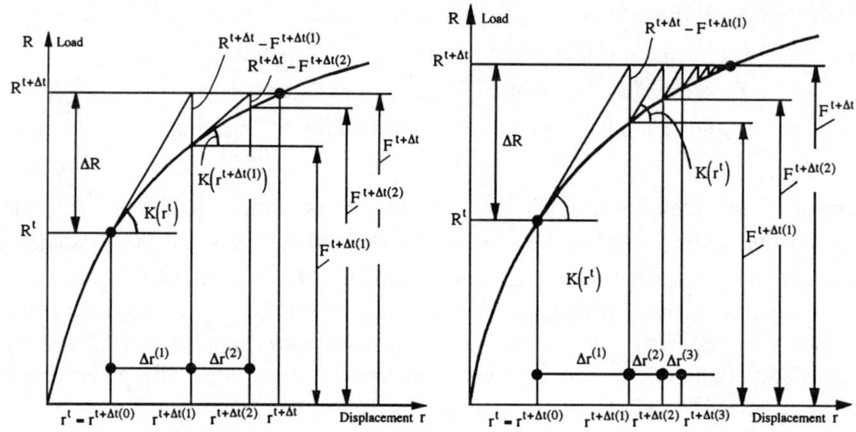

Fig. 3.5. Full and modified Newton–Raphson methods
(reprinted from (Kleiber 89) with his kind permission)

Modified Newton-Raphson Scheme. (3.86) defines the full Newton–Raphson iteration scheme, in which the updating and factorizing of the effective stiffness matrix takes place in each iteration (Fig. 3.5). Though it can be efficient in some specific non-linear analyses, the use of the full Newton–Raphson scheme is usually not very efficient in general geometric and material non-linear response calculations. It is then better to use a *modified Newton–Raphson* scheme as:

$$\mathbf{M}_n \ddot{\mathbf{U}}_{n+1\_i} + \mathbf{D}_n \dot{\mathbf{U}}_{n+1\_i} + \mathbf{K}_n \Delta \mathbf{U}_i = \mathbf{L}_{n+1} - \mathbf{R}_{n+1\_i-1} \tag{3.88}$$

in which the tangent stiffness matrix $\mathbf{K}_n$ is not updated at each iteration, in order to factorize the effective stiffness matrix only once at each time step. All non-linearities are then fully taken into account in the evaluation of the vector $\mathbf{R}_{n+1\_i-1}$. Clearly, the number of algebraic operations required in each case to reach the convergence may differ considerably (Kleiber 89).

3.3.4 The Modal Superposition Method

Generally, it may be said that the number of operations required in direct integration methods is directly proportional to the number of time step used in the analysis. Therefore, explicit integration can be expected to be effective when a response for a relatively short duration is required. However, if the integration must be carried out for many time steps, it may be more effective first to transform the equilibrium equation into a form in which the step-by-step solution is less costly. The basic operation in this solution is a change of basis from p generalized *nodal* displacements to q generalized *modal* displacements $(q << p)$, prior to the incremental resolution (Kleiber 89). For this purpose, the generalized nodal displacement vector may be expressed in terms of the generalized modal displacement vector as:

$$\mathbf{U}_n = \boldsymbol{\Psi}\,\mathbf{V}_n \qquad \text{with} \qquad \boldsymbol{\Psi}_{p\times q} = \left[\boldsymbol{\psi}_1 \;\cdots\; \boldsymbol{\psi}_q\right] \tag{3.89}$$

where $\mathbf{V}_n$ is the generalized modal displacement vector at step n
$\boldsymbol{\psi}_k$ are the eigenvectors of the linearized eigenproblem at time 0
$\boldsymbol{\Psi}$ is the modal transfer composed of the column eigenvectors, as:

$$\mathbf{K}_0\boldsymbol{\Psi} = \mathbf{M}_0\boldsymbol{\Psi}\,\boldsymbol{\Omega}^2 \qquad \text{with} \qquad \boldsymbol{\Omega}^2 = \begin{bmatrix} \omega_1^2 & 0 & 0 \\ 0 & \ddots & 0 \\ 0 & 0 & \omega_q^2 \end{bmatrix} \qquad \omega_k = 2\pi f_k \tag{3.90}$$

where the f_k are the natural frequencies of the system $\left(k \equiv (1,\cdots,q)\right)$ (Kleiber 89). Applying (3.90) at time $n+1$ and iteration i in (3.76) leads to the modal equation:

$$\tilde{\mathbf{M}}_n\ddot{\mathbf{V}}_{n+1\_i} + \tilde{\mathbf{D}}_n\dot{\mathbf{V}}_{n+1\_i} + \tilde{\mathbf{K}}_n\Delta\mathbf{V}_i = \tilde{\mathbf{L}}_{n+1} - \tilde{\mathbf{R}}_{n+1\_i-1} \qquad \text{with} \tag{3.91}$$

$$\tilde{\mathbf{M}}_n = \boldsymbol{\Psi}^T\mathbf{M}_n\boldsymbol{\Psi} \qquad \tilde{\mathbf{D}}_n = \boldsymbol{\Psi}^T\mathbf{D}_n\boldsymbol{\Psi} \qquad \tilde{\mathbf{K}}_n = \boldsymbol{\Psi}^T\mathbf{K}_n\boldsymbol{\Psi}$$

$$\tilde{\mathbf{L}}_{n+1} = \boldsymbol{\Psi}^T\mathbf{L}_{n+1} \qquad \tilde{\mathbf{R}}_{n+1\_i-1} = \boldsymbol{\Psi}^T\mathbf{R}_{n+1\_i-1}$$

where $\tilde{\mathbf{M}}_n$, $\tilde{\mathbf{D}}_n$, $\tilde{\mathbf{K}}_n$ are the modal tangent mass, damping, stiffness matrix
$\tilde{\mathbf{R}}_{n+1\_i-1}$ and $\tilde{\mathbf{L}}_{n+1}$ are the modal internal and external force vectors

As the eigenvectors are $\mathbf{M}$-orthogonal, (3.91) may be written in the usual form:

$$\ddot{\mathbf{V}}_{n+1\_i} + \boldsymbol{\Lambda}\dot{\mathbf{V}}_{n+1\_i} + \boldsymbol{\Omega}^2\Delta\mathbf{V}_i = \tilde{\mathbf{L}}_{n+1} - \tilde{\mathbf{R}}_{n+1\_i-1} \qquad \text{with} \tag{3.92}$$

$$\tilde{\mathbf{M}}_n \equiv \mathbf{I} \qquad \tilde{\mathbf{D}}_n = \boldsymbol{\Lambda} \qquad \tilde{\mathbf{K}}_n = \boldsymbol{\Omega}^2$$

This method may of course be efficiently applied for linear dynamics analysis, since in the linear case, (3.92) is a linear diagonal system. By contrary, in the general case, the incremental equilibrium equations defined in the new basis by (3.92) are still coupled because the nodal internal force vector $\mathbf{R}_{n+1\_i-1}$ can only be evaluated when all displacements are known. The solution of the eigenproblem defined by (3.90), added to the step-by-step solution of (3.88), using for instance the Newmark method, may prove out significantly more cost-effective than the direct step-by-step solution of (3.88). However, the numerical gain depends on how significant are the non-linearities in the system (Kleiber 89).

Conclusion

Linear analyses usually lead to one exact solution, whereas non-linear ones lead to slightly different approximate solutions depending on the incremental method applied. As shown in this chapter, various numerical methods have been developed for solving non-linear algebraic systems. The finite element method is applied for geometric discretization, while the finite difference and linear iteration methods are invoked for temporal integration. Various incremental / iterative algorithms may then be applied to solve the linearized incremental systems. Usually iterations are applied within the increment steps to reduce the errors induced by the finite approximations. There is not a unique solution method, rather a set of possible approaches, whose efficiency depends on the problem to solve. For soft tissue modeling, it may therefore be interesting to distinguish the different kinds of available biomechanical models to apprehend the way these incremental methods may be applied. This is the purpose of the next chapter.

References

This chapter is essentially based on (Trompette 92), (Hughes 87), (Katona 78, 80), (Kleiber 89), (Argyris 80), (Desai 80), (Lee 83), (CHARM D4). For deeper investigation, refer also to (Bathe 96), (Crisfield 91), (Curnier 94), (Desai 90), (Oden 89), (Distefano 74), (Simo 87), (Zienkiewicz 89, 91).

Argyris 80 J.H Argyris, J.St. Doltsinis, W.C. Knudson, J. Szimmat, K.J. Willam, H. Wüstenberg (1980), Eulerian and Lagrangian techniques for elastic and inelastic large deformation processes, in *Computational Methods in Non-linear Mechanics*, ed. by J.T. Oden, E.B.Becker. Amsterdam: North-Holland Publishing Co.

Bathe 96 K.-J. Bathe (1996), Finite element procedures, Review of: *Finite Element Procedures in Engineering Analysis*, 1982, Englewood Cliffs, NJ: Prentice Hall

CHARM D4 W. Maurel, Y. Wu (1994), Survey of mechanical models for tissue deformation and muscle contraction with identification of parametric data, LIG-EPFL/MIRALab-UG, ESPRIT 9036 Project CHARM, Internal Deliverable D4

Crisfield 91 M.A. Crisfield (1991), *Non-linear finite element analysis of solids and structures,* Chichester, UK: John Wiley & Sons

Curnier 94 A. Curnier (1994), *Computational methods in solid mechanics*, Dordrecht: Kluwer Academic Publishers

Desai 80 C.S. Desai, H.V. Phan (1980), Three-Dimensional Finite Element Analysis including Material and Geometric Non-linearities, in *Computational Methods in Non-linear Mechanics*, ed. by J.T. Oden, E.B. Becker. Amsterdam: North-Holland Publishing Co.

Desai 90 C.S. Desai, G. Gioda (eds.) (1990), *Numerical Methods and Constitutive Modelling in Geomechanics.* Vienna: Springer

Distefano 74 N. Distefano (1974), *Non-linear Processes in Engineering.* New York: Academic Press

Hughes 87 T.J.R. Hughes (1987), *The Finite Element Method: linear static and Dynamic Finite Element Analysis,* Englewood Cliffs, NJ: Prentice-Hall

Katona 78 M.G. Katona (1978), *A viscoelastic-plastic constitutive model with a finite element solution methodology: technical final report*, 1974–1977; prep.: Civil Engineering Laboratory, Naval Construction Battalion Center, Port Hueneme–California 93043

Katona 80 M.G. Katona (1980), Combo-viscoplasticity: an introduction with incremental formulation, *Computers & Structures*, 11, 217–224

Kleiber 89 M. Kleiber (1989), *Incremental Finite Element Modeling in Non-linear Solid Mechanics*, transl. ed.: C. T. F. Ross. Chichester: Ellis Horwood

Lee 83 G.C. Lee, N.T. Tseng, Y.M. Yuan (1983), Finite element modeling of lungs including interlobar fissures and the heart cavity, *J. Biomechanics*, 16, 9, 679–690

Oden 89 J.T. Oden, J.M. Steele (1989), *Finite Elements of Non-linear Continua*, New York: Dekker

Simo 87 J.C. Simo (1987), On a fully three-dimensional finite-strain viscoelastic damage model: formulation and computational aspects, *Computer Methods in Applied Mechanics and Engineering*, 60, 153–173

Trompette 92 P. Trompette (1992), *Mécanique des structures par la méthode des éléments finis: Statique et Dynamique*. Paris: Masson

Zienkiewicz 89 O.C. Zienkiewicz, R.L. Taylor (1989), *The finite element method,* 4th edn. London: McGraw-Hill

Zienkiewicz 91 O.C. Zienkiewicz, R.L. Taylor (1991), *Solid and Fluid Mechanics, Dynamics and Non-linearity*, 4th edn. London: McGraw-Hill

4 Constitutive Modeling

The objective of experimentation is to obtain simple, general laws describing the macroscopic behavior of materials, in order to determine their mechanical properties, and predict their response under well defined conditions (Fung 87). However, due to a lack of measurement methods and data, the characterization of living materials can hardly be completely assessed (Lee 82). Therefore, more than elaborating an extensive review on the topic, this chapter aims at presenting the various approaches for soft tissue modeling, and outlining the general forms of the available models in order to apprehend their suitability for soft tissues deformation simulation.

4.1 Phenomenological Modeling

4.1.1 Principle

Phenomenological modeling mainly consists of fitting mathematical equations to experimental tensile curves. This approach is convenient for generalization, and behavior prediction in independent tests (Woo 93). In the following, T represents the Lagrange stress (2.43), ε the infinitesimal strain (2.28), and λ the corresponding extension ratio (2.16) defined by:

$$\lambda = 1 + \varepsilon \tag{4.1}$$

Various stress-strain relationships have been derived from curves, fitting experimental data or based on the continuum mechanics:

$$\varepsilon^2 = a\,T^2 + b\,T \qquad \text{for tendons} \qquad \text{(Wertheim 47)} \tag{4.2}$$

$$\begin{cases} \varepsilon = a\, T^b \\ T = a\, \varepsilon^b \end{cases}$$ for collagen fibers (Morgan 60) (4.3)

$$\varepsilon = a\left(1 - e^{bT}\right)$$ for elastic fibers (Carton 62) (4.4)

$$T = a\, \varepsilon^b$$ for rat tail skin (Kenedi 64) (4.5)

$$T = a\, \varepsilon^b$$ for guinea pig skin (Glaser 65) (4.6)

$$\varepsilon = a\, T^b + c$$ for skin (Ridge 66) (4.7)

$$T = a\, \varepsilon^b$$ for skin (Elden 68) (4.8)

Vlasblom applied the Hooke law to simulate skin deformation at low strain. Torsion tests were performed on vivo human arm skin for wrist angles less than 0.03 rad (Vlasblom 67). A linear relation was also assumed at low loads by Sakata research for in vitro sheep abdominal skin (Sakata 72). Elden, the first, proposed a model with ring uniaxial tests on rat tail skin. The difference between experiment data and the extrapolated linear portion was obtained as shown in Fig. 4.1.

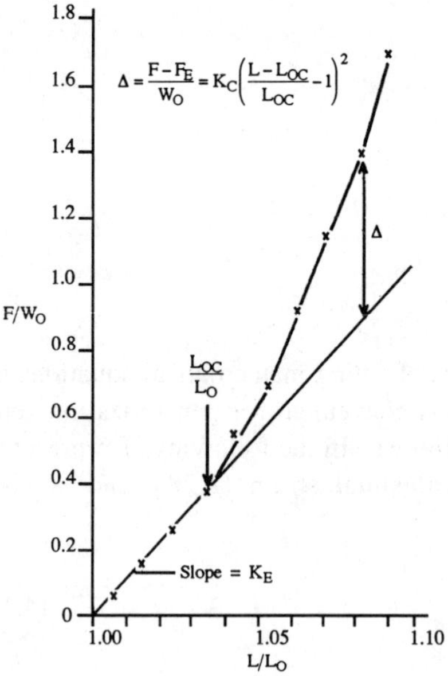

Fig. 4.1. Typical dependence of overall stress on strain

(reprinted from (Elden 68) with permission of John Wiley & Sons)

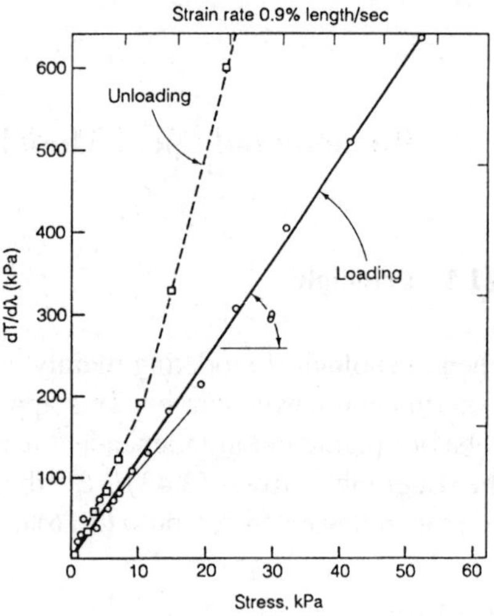

Fig. 4.2. Load versus load-extension ratio curve for soft tissues

(reprinted from (Fung 72) with permissions of Prentice- Hall)

4.1.2 The Quasi-Linear Viscoelasticity Theory

According to Fung, it is reasonable to expect that for oscillations of small amplitude about an equilibrium state, the theory of linear viscoelasticity should apply. For finite deformations, however, the non-linear stress-strain characteristics of the living tissues must be taken into account (Fung 72). Thus, he has developed, for soft tissues, a theory he has called *quasi-linear viscoelasticity* (QLV), by assuming the *relaxation function* $K(\lambda, t)$ as composed of a *reduced relaxation function* $G(t)$ and an *elastic response* $T^{(e)}(\lambda)$, such as:

$$K(\lambda, t) = G(t)\, T^{(e)}(\lambda) \qquad \text{leading to the Lagrangian stress in the form:} \qquad (4.9)$$

$$T(t) = \int_{-\infty}^{t} G(t-\tau) \frac{\partial T^{(e)}[\lambda(\tau)]}{\partial \tau}\, d\tau \qquad \text{with} \qquad G(0) = 1 \qquad (4.10)$$

which may be seen as the sum of the contributions of all the past changes, each governed by the same reduced relaxation function $G(t)$ (hence the name *quasi-linear viscoelasticity*). The QLV theory may be equivalently formulated for creep with a *reduced creep function* $J(t)$, as:

$$\varepsilon(T, t) = \int_{-\infty}^{t} J(t-\tau) \frac{d\varepsilon^{(e)}[T(\tau)]}{d\tau}\, d\tau \qquad (4.11)$$

The forms recommended by Fung for the elastic response $T^{(e)}(\lambda)$, the reduced relaxation $G(t)$ and creep $J(t)$ functions are presented in the following.

4.1.3 The Elastic Response

Assuming $T = \varepsilon = 0$ for $t < 0$ and rewriting (4.10) in the form:

$$T(t) = T^{(e)}[\lambda(t)] + \int_{0}^{t} T^{(e)}[\lambda(t-\tau)] \frac{\partial G(\tau)}{\partial \tau}\, d\tau \qquad (4.12)$$

Fung has noticed that, since the steady-state stress-strain relationships of soft tissues are not very sensitive to strain rate, their elastic response $T^{(e)}(\lambda)$ may be described using elasticity theory – a property he has called *pseudoelasticity* – and may be approximated by their tensile stress response in a loading experiment with a sufficiently high rate of loading. Furthermore observing that, after preconditioning, their stress-strain loop is repeatable, he has pointed out that the loading and unloading tensile curves may be characterized separately using the elasticity theory (Fung 72). The QLV theory may thus be applied using any stress-strain relationship as elastic response.

The Exponential Form. Fung, the first, outlined a linear relationship between the stress T and the slope $dT/d\lambda$ of the stress-extension curve of soft tissues as:

$$\frac{dT}{d\lambda} = a(T+b) \qquad \text{leading to} \qquad (4.13)$$

$$T = (T_0 + b)\, e^{a(\lambda-\lambda_0)} - b \qquad \text{with} \qquad b = \frac{P_0\, e^{-a\lambda_0}}{1-e^{-a\lambda_0}} \qquad (4.14)$$

for strains up to about 30% resting length, where (λ_0, T_0) is an arbitrary point on the curve (Fig. 4.2) (Fung 67). This relation has been applied to the determination of rabbit papillary muscle properties at various temperatures and strain rates (Pinto 1973). For larger strains, Fung recommended to use the form:

$$T = a\left(\lambda - \frac{1}{\lambda^2}\right) e^{b\lambda} \qquad (4.15)$$

For rabbit mesentery, Fung obtained a better fit with a parabola as (Fung 67):

$$\frac{dT}{d\lambda} = aT\,(1-bT) \qquad \text{leading to} \qquad (4.16)$$

$$T = \frac{e^{a\lambda}}{c + be^{a\lambda}} \qquad \text{with} \qquad c = \left(\frac{1}{T_0} - b\right) e^{a\lambda_0} \qquad (4.17)$$

Fung's result has led to various similar formulations in terms of strain ε such as:

$$\frac{dT}{d\varepsilon} = aT + b \qquad \text{for skin} \qquad \text{(Wijn 76)} \qquad (4.18)$$

$$\begin{cases} T = a\left(e^{b\varepsilon} - 1\right) & \text{for ventricle muscle fibers} \quad \text{(Nevo 89)} \\ \begin{cases} T_c = \bar{a}\left(e^{\bar{b}(\varepsilon_c-\varepsilon_r)} - 1\right) & \varepsilon_c > \varepsilon_r \\ T_c = 0 & \varepsilon_c \le \varepsilon_r \end{cases} & \begin{cases} \varepsilon_c \text{ the extension ratio of collagen fibers} \\ \varepsilon_r \text{ the extension ratio where collagen stress begins} \end{cases} \end{cases} \qquad (4.19)$$

$$T = a\left[e^{b\varepsilon} - 1\right]$$ for articular cartilage (Woo 76), for the heart muscle (Feit 79) (4.20)

and to simplified or modified formulations accounting for specific properties, as:

$$T = a\left(\varepsilon - \varepsilon^2\right) e^{b\varepsilon}$$ for anterior cruciate ligaments (Haut 69) (4.21)

$$T = a\,\varepsilon + b\,\varepsilon^2$$ for elastic tissues (Haut 72) (4.22)

$$T = \frac{E}{2}\left[\lambda\left(\lambda^2 - 1\right)\right]$$ for left ventricle (Chadwick 81) (4.23)

$$T = \left(a\dot{\varepsilon} + b\right)\left(e^{c\varepsilon} - 1\right)$$ at high strain rate $\left(\dot{\varepsilon} < 3.5 s^{-1}\right)$ for human aortic tissue (Collins 72) (4.24)

4.1.4 The Reduced Relaxation and Creep Functions

Stress Relaxation. Observing the fact that the stress-strain loop is almost independent of the strain rate with several decades of the rate variation, Fung has suggested applying a reduced relaxation function in the form (Fung 72):

$$G(t) = \frac{1 + \int_0^\infty S(\tau) e^{-\frac{t}{\tau}} d\tau}{1 + \int_0^\infty S(\tau)\, d\tau} \qquad S(t) = \begin{cases} \frac{c}{t} + S_0 & \tau_1 \le t \le \tau_2 \\ 0 & t < \tau_1, t > \tau_2 \end{cases} \qquad \text{i.e.:} \quad (4.25)$$

$$G(t) = \frac{1 + c\left[E\left(\frac{t}{\tau_2}\right) - E\left(\frac{t}{\tau_1}\right)\right]}{1 + c\ln\left(\frac{\tau_2}{\tau_1}\right)} \qquad E(y_1) = \int_{y_1}^\infty \frac{e^{-y}}{y}\, dy \qquad \begin{cases} y = \frac{t}{\tau} \\ y_1 = \frac{t}{\tau_1} \end{cases} \qquad \text{i.e.:} \quad (4.26)$$

$$G(t) \approx \frac{1 - c\,\gamma - c\ln\left(\frac{t}{\tau_2}\right)}{1 + c\ln\left(\frac{\tau_2}{\tau_1}\right)} \qquad \gamma = 0.5772 \qquad \tau_1 << t << \tau_2 \qquad \text{i.e.:} \quad (4.27)$$

$$G(t) = \alpha \ln(t) + \beta \qquad \alpha = -\frac{c}{1 + c\ln\left(\frac{\tau_2}{\tau_1}\right)} \qquad \beta = \frac{1 - c\,\gamma + c\ln(\tau_2)}{1 + c\ln\left(\frac{\tau_2}{\tau_1}\right)} \tag{4.28}$$

Many authors have applied the QLV theory in combining an elastic response as defined by (4.19) with a reduced relaxation function as defined by (4.28), i.e.:

$$T^{(e)} = a\left(e^{b\varepsilon} - 1\right) \qquad G(t) = \alpha \ln(t) + \beta \quad \text{with} \quad \alpha(c, \tau_1, \tau_2) \quad \text{and} \quad \beta(c, \tau_1, \tau_2)$$

Woo applied it to model the canine medial collateral ligament (Woo 82), Trevisan to model lamb and human tendons and ligaments (Trevisan 83), Kwan et al. to model the anterior cruciate ligament (Kwan 93). Best et al. used (4.28) for rabbit live skeletal muscle modeling (Best 94). The QLV theory has also been applied by Lin et al. for porcine anterior cruciate ligament modeling (Lin 87), and by Lyon et al. for human anterior cruciate ligament and patellar tendon modeling (Lyon 88).

Some authors have modified the reduced relaxation function to account for specific properties. Haut and Little have noted that the stress relaxation response is dependent on the strain level ε_0 (Haut 72). For stress relaxation experiments on collagen fiber bundles extracted from rat tail tendons, they have used:

$$T = a\,\varepsilon^b \qquad \text{and} \qquad G(\varepsilon_0, t) = E\,\varepsilon_0^2\,[1 + \mu \ln(t)] \tag{4.29}$$

Similarly, Jenkins and Little have noticed that the slope of the stress relaxation curves is dependent upon the square of the strain level ε_0 (Jenkins 74). For stress relaxation experiments on bovine ligamentum nuchae, they have used:

$$T = a\,\varepsilon + b\,\varepsilon^2 \qquad \text{and} \qquad G(\varepsilon_0, t) = 1 + \mu\,\varepsilon_0^2(t)\ln(t) \tag{4.30}$$

Creep. The reduced creep function of (4.11) may be similarly expressed as:

$$J(t) = \frac{1 - c\left[E\left(\frac{t}{c + \tau_2}\right) - E\left(\frac{t}{c + \tau_1}\right)\right]}{1 - c\ln\left(\frac{c + \tau_2}{c + \tau_1}\right)} \qquad \text{or} \qquad J(t) = \alpha \ln(t) + \beta \tag{4.31}$$

Galford and McElhaney applied it to creep experiments on scalp, brain and dura from both monkey and human (Galford 70). Pradas and Calleja also used it to model the crimp behavior of human flexor tendons with an elastic response as:

$$\begin{cases} \varepsilon^{(e)}(T) = A \ln\left(BT + \sqrt{(BT)^2 + 1}\right) & 0 \le T \le T_0 \\ \varepsilon^{(e)}(T) = \dfrac{T - d}{m} & T_0 \le T \le T_{max} \end{cases} \quad \text{(Pradas 90)} \qquad (4.32)$$

The valuable interest of the QLV theory is that it allows to characterize separately the purely elastic response and the time dependent behavior of the soft tissues. The QLV assumption has thus been widely applied to soft tissue characterization.

4.2 Structural Modeling

4.2.1 Uniaxial Elastic Models

Structural models are based on the assumed behavior of the structural components of the tissue. They are thus formulated in terms of the structural parameters of the tissue. They appear more appropriate for relating the microstructure of the tissues to their mechanical behavior (Woo 93).

Diamant et al. developed a bent elastica model for the crimped collagen fibers of tendon, in the form of series of slender crimp arms medially connected by springs (Fig. 4.3). The resulting mechanical behavior was described by:

$$\varepsilon - \frac{T}{E} = \varepsilon_\infty - \gamma\sqrt{\frac{E}{T}} \quad \text{with} \qquad (4.33)$$

$$\gamma = \frac{D}{l}\,\frac{1-\cos\left(\frac{1}{2}\theta_0\right)}{\cos\left(\theta_0\right)} \qquad l = l_0 \sec\left(\theta_0\right) \qquad \varepsilon_\infty = \sec\left(\theta_0\right) - 1$$

where ε is the macroscopic strain and T the macroscopic stress
ε_∞ the strain, θ_0 the crimp angle
l_0 the half period of the unstretched fiber
E the elastic modulus and D the diameter of the fiber

Geometrical parameters have been collected on mature rat tendons specimens for validation of the established relation (Diamant 72).

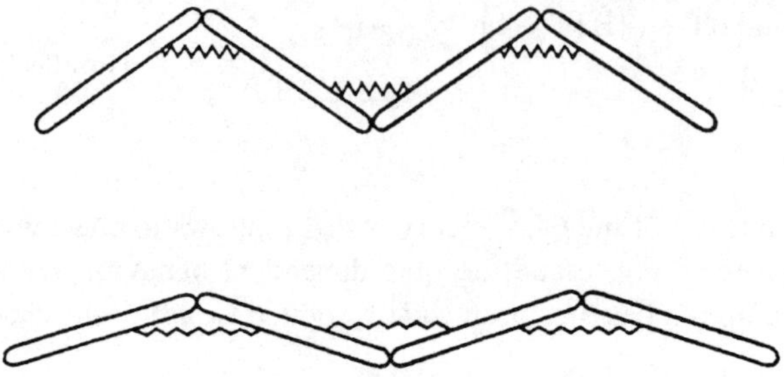

Fig. 4.3. Diamant's bent elastica model
(reprinted from (Diamant 72) with permission of The Royal Society)

Purslow applied this model to predict the mechanical behavior of bovine muscle perimysium in formulating the elastic modulus E_L of the collagenous fiber network along the muscle fiber direction (Fig. 4.4) as (Purslow 89):

$$E_L = \frac{dT}{d\varepsilon} = E_f \, V_f \cos^4(\alpha) \qquad \text{with} \qquad (4.34)$$

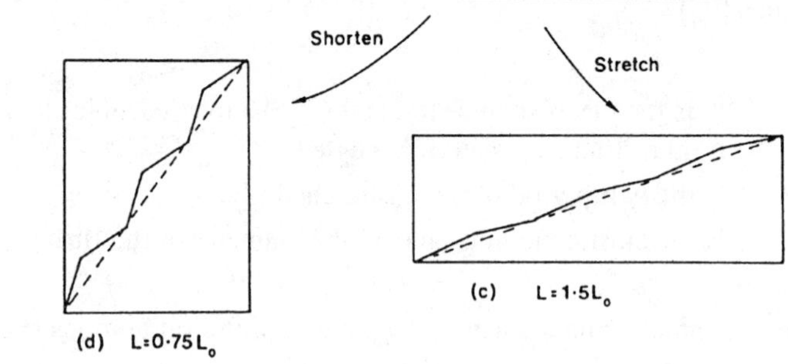

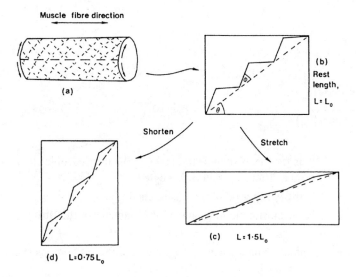

Fig. 4.4. Purslow's muscle model
(reprinted from (Purslow 89) with permission of Elsevier Science)

$$E_f = \frac{dT_f}{d\varepsilon_f} = \left(\frac{1}{E} + \frac{\gamma}{2T_f} \sqrt{\frac{E}{T_f}} \right)^{-1} \tag{4.35}$$

and

α	mean collagen fiber angle with respect to the muscle fiber
T_f, ε_f	stress and strain of an individual fiber
E_f, V_f	axial elastic modulus and volume fraction of fibers

α is known from the geometric model as a function of the sarcomere length. The model has been applied to the characterization of the bovine perimysium (Purslow 89).

Beskos and Jenkins described the capability of soft tissues to adjust themselves to the finite changes in size and shape without affecting the *tensile strain*, with help of a hollow cylinder, braided with inextensible fibers arranged in a helical pattern (Fig. 4.5) (Beskos 75).

Fig. 4.5. A braided hollow cylinder conformation for tissue modeling (reprinted from (Crisp 72) with permission of Prentice-Hall)

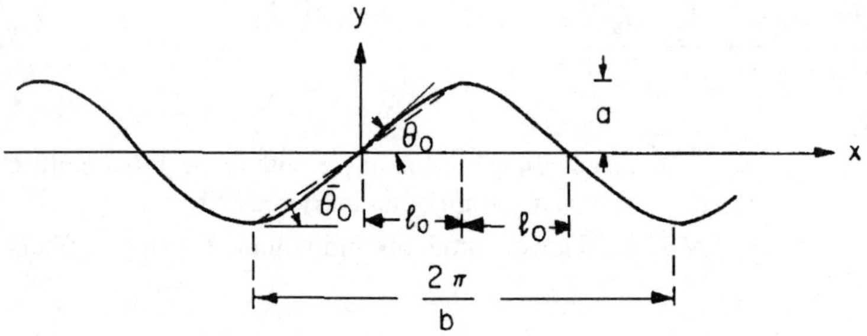

Fig. 4.6. Comninou's model for collagen fibers
(reprinted from (Comninou 76) with permission of Elsevier Science)

Comninou and Yannas assumed the non-linear behavior of soft tissues due to the waviness of the fibers and modeled them as sinusoidal beams (Fig. 4.6) (Comninou 76). The macroscopic strain $\bar{\varepsilon}$ has then been expressed as:

$$\bar{\varepsilon} = \varepsilon + \frac{a^2 b^2}{4} \frac{\Lambda(\Lambda+2)}{(\Lambda+1)^2} \quad \text{with} \tag{4.36}$$

$$\Lambda = \frac{4\,\varepsilon}{b^2\,R^2} \qquad \varepsilon = \frac{T}{E} \qquad b = \frac{\pi}{2l_0} \qquad R^2 = \frac{4I}{A} = \frac{4D}{EA} \qquad a = \frac{1}{b}\tan(\theta_0)$$

and T as Lagrangian stress, E tensile stiffness, **I** bending rigidity,
D as diameter and A as cross-sectional area of the fiber
a as wave amplitude, l_0 as wave half period of unstretched fiber

The model is restricted to small uniaxial strains and constant crimp configuration. Using this model, curves have been plotted and compared with those obtained by Diamant et al. (Diamant 72).

Kastelic et al. introduced a model in which the resistance to deformation arises only from the elasticity of already straightened collagen fibrils (Fig. 4.7). The stress was described by:

$$T = E\,e(R)^2 - 2(1-b)E \int_0^R \left(\frac{1}{\cos\theta(\rho)} - 1 \right) \rho \, d\rho \qquad \text{where} \tag{4.37}$$

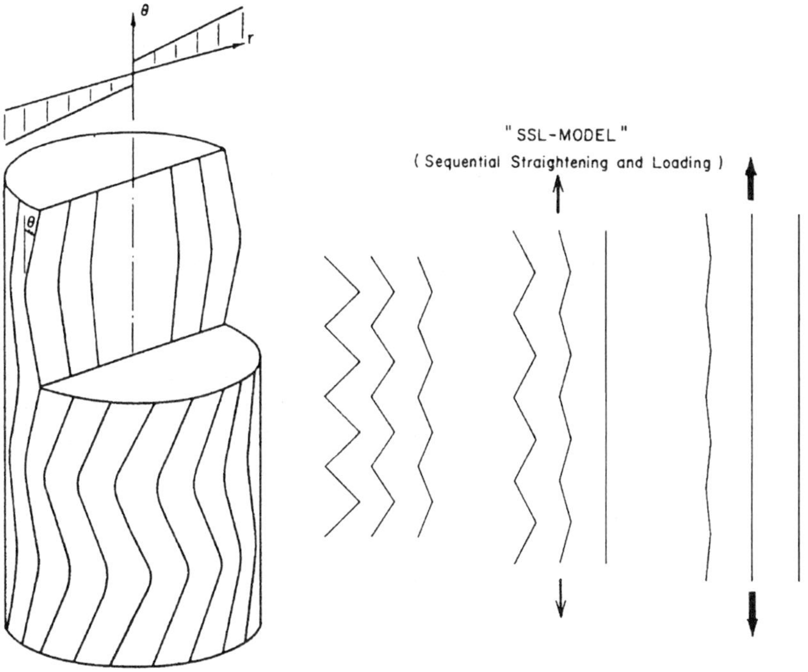

Fig. 4.7. Kastelic's model for collagen fibers
(reprinted from (Kastelic 80) with permission of Elsevier Science)

$$R = \left[\frac{1}{1-a}\left(\frac{1}{\theta_0}\arccos\left(1+\frac{\varepsilon}{1-b}\right)^{-1} - a\right)\right]^{\frac{1}{p}} \qquad (4.38)$$

is the relative radius of the internal circular area containing straightened fibrils,

$$e(\rho) = (1-b)\left(\frac{1}{\cos\left[\theta(\rho)\right]} - 1\right) \qquad (4.39)$$

is the relative strain of a fibril before complete straightening,

$$\theta(\rho) = \theta_0\left[a + (1-a)\,\rho^p\right] \qquad (4.40)$$

is the distribution of the initial crimp angles of the fibers across the radius ρ, and

θ_0 is the initial crimp angle of the outermost fiber
a and p are the factors of the crimp angle distribution
ρ is the relative radius of the fiber position in the beam ($0 \le \rho \le 1$),
E is the fiber Young modulus
b is the crimp blunting factor
ε is the relative strain of the fascicle

Curves have been plotted for a set of values and compared to uniaxial tensile characteristics of rat tail tendons (Kastelic 80).

Decraemer et al. assumed soft tissues constituted of purely elastic fibers, with same moduli and cross-sectional areas, embedded in a gelatin-like liquid, and more or less folded because of different lengths (Fig. 4.8) (Decraemer 80a). The resulting behavior was described by the quasi-elastic stress-strain relation:

$$T(l) = \frac{1}{\sqrt{2\pi}} \frac{b}{\mu s} \left[l \int_{l_0}^{l} e^{-\frac{(\mu - l_i)^2}{2s^2}} dl_i - \int_{l_0}^{l} l_i e^{-\frac{(\mu - l_i)^2}{2s^2}} dl_i \right] \quad \text{with} \quad b = \frac{k N a}{A_0} \tag{4.41}$$

where: l and l_0 are the actual and initial lengths of the test specimen
A_0 its initial cross-section, b its effective Young's modulus
N the number of fibers i, k their Young's modulus
a their average cross-section, l_i is their initial lengths
μ and s, mean value and standard deviation of length distribution

The model has been fitted to uniaxial tensile curves of human vein, fascia, tympanic membrane, and rabbit papillary muscle (Decraemer 80a).

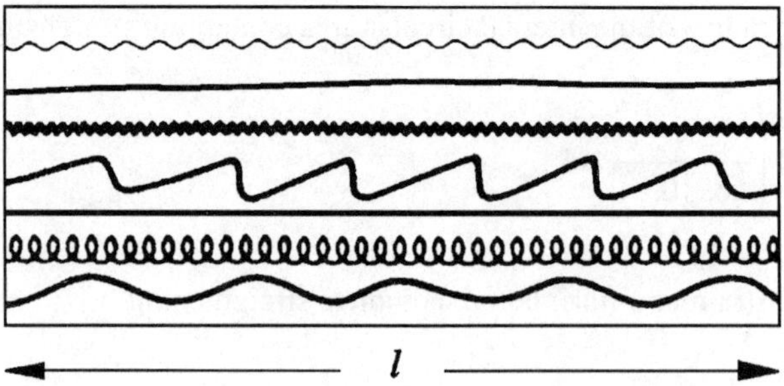

Fig. 4.8. Decraemer's model for soft tissues
(reprinted from (Decraemer 80a) with permission of Elsevier Science)

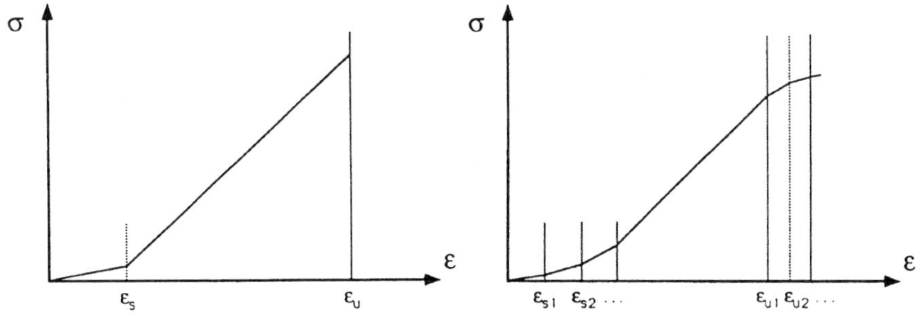

Fig. 4.9. Kwan's model for soft tissues
(reprinted from (Kwan 89) with permission of ASME)

Kwan and Woo proposed a fibered model, in which the behaviors of the fibers are described by bilinear tensile curves with different slopes for the toe and linear regions (Fig. 4.9), i.e., with stress-strain relationships of the form:

$$T = \begin{cases} E_S \varepsilon & 0 < \varepsilon \le \varepsilon_S \\ E_e(\varepsilon - \varepsilon_S) + E_S \varepsilon_S & \varepsilon_S < \varepsilon < \varepsilon_U \end{cases} \tag{4.42}$$

In the model, the fibers are assumed to belong to m groups differing on their lengths, and n groups differing on their failure strain limits. A constitutive relationship has then been formulated for a 3×3 model as follows: (4.43)

$$T = \begin{cases} E_S\, \varepsilon & 0 < \varepsilon \le \varepsilon_{S1} \\ \left[E_S + \gamma_1(E_e - E_S)\right]\varepsilon - (E_e - E_S)\,\gamma_1\varepsilon_{S1} & \varepsilon_{S1} < \varepsilon \le \varepsilon_{S2} \\ \left[E_S + (\gamma_1 + \gamma_2)\,(E_e - E_S)\right]\varepsilon - (E_e - E_S)\,(\gamma_1\varepsilon_{S1} + \gamma_2\varepsilon_{S2}) & \varepsilon_{S2} < \varepsilon \le \varepsilon_{S3} \\ E_e\,\varepsilon - (E_e - E_S)\,(\gamma_1\varepsilon_{S1} + \gamma_2\varepsilon_{S2} + \gamma_3\varepsilon_{S3}) & \varepsilon_{S3} < \varepsilon \le \varepsilon_{U1} \\ (1 - \beta_1)\,E_e\,\varepsilon - (E_e - E_S)\,(\gamma_1\varepsilon_{S1} + \gamma_2\varepsilon_{S2} + \gamma_3\varepsilon_{S3}) + \beta_1\varepsilon_{U1}E_e & \varepsilon_{U1} < \varepsilon \le \varepsilon_{U2} \\ \beta_3 E_e\,\varepsilon - (E_e - E_S)\,(\gamma_1\varepsilon_{S1} + \gamma_2\varepsilon_{S2} + \gamma_3\varepsilon_{S3}) + (\beta_1\varepsilon_{U1} + \beta_2\varepsilon_{U2})E_e & \varepsilon_{U2} < \varepsilon \le \varepsilon_{U3} \end{cases}$$

It has been applied to rabbit anterior cruciate and canine medial collateral ligaments modeling (Kwan 89).

4.2.2 Uniaxial Viscoelastic Models

Viscoelasticity may be modeled by adding viscoelastic properties to previous fibrous elastic models. This approach was followed by Decraemer *et al.* who added internal friction between the fibers and the matrix (Decraemer 80b). The uniaxial elastic stress defined in (4.41) then became viscoelastic in the form:

$$T(t) = \int_{-\infty}^{t} G_r(t-\tau)\frac{dT^{(e)}[l(\tau)]}{d\tau}\,d\tau \quad \text{with a reduced relaxation function as:} \qquad (4.44)$$

$$G_r(t) = \frac{G(t)}{G(0)} = G_r^{\infty} + \int_0^{\infty} H_r(\tau)\, e^{-\frac{t}{\tau}}\, d[\ln(\tau)] \qquad \text{with} \qquad (4.45)$$

$$H_r(t) = \left[1 + c\ln\frac{\tau_2}{\tau_1}\right]^{-1} \qquad \text{for} \quad \tau_1 \le t \le \tau_2 \quad \text{and} \qquad (4.46)$$

$$T^{(e)}(t) = \frac{1}{\sqrt{2\pi}}\frac{N}{s}\frac{G(0)}{A_0}\int_{l_0}^{l(t)} \frac{l(t)-l_i}{l_i}\, e^{-\frac{(\mu-l_i)^2}{2s^2}}\, dl_i \qquad \text{as elastic response} \qquad (4.47)$$

The determination of $G_r(t)$ has thus here been replaced by the determination of $H_r(t)$ for human tympanic membrane with harmonic strain oscillations (Decraemer 80b). A final improvement was brought by Maes et al. who assumed fibers composed of a large number of zigzag elements with cross-links and included thermodynamic relationships as (Maes 89):

$$T(\lambda,\theta) = \frac{1}{\sqrt{2\pi}}\frac{b(\theta)}{s'}\int_0^{\lambda} \frac{\lambda-\lambda_i}{\lambda_i}\, e^{-\frac{(\lambda_i-\mu')^2}{2s'^2}}\, d\lambda_i \qquad \text{with} \qquad (4.48)$$

$$f(\theta) = \frac{k(\theta-\theta_0)(1+\alpha)+Cd(1-\alpha)^2}{2\alpha k(\theta-\theta_0)+Cd(1-\alpha)^2} \qquad b(\theta) = \frac{N}{A_0}\left[\frac{k(1+\alpha)(\theta-\theta_0)+Cd(1-\alpha)^2}{d(1-\alpha)^2}\right]$$

$$\mu' = \frac{\mu^0}{l_0^0} \quad s' = \frac{s^0}{l_0^0} \quad \lambda_i = \frac{l_i^0}{l_0^0} \qquad \lambda = \frac{1}{l_0(\theta)} \qquad l_i(\theta) = l_i^0 f(\theta) \qquad \alpha = \cos(\beta/2)$$

where
- d is length of a zigzag element in the fiber model
- β is the angle between two zigzag elements
- l_i^0 is the length of a fiber when no cross-link is broken
- θ_0 is the temperature below which no cross-link is broken
- μ^0, s^0 and l_0^0 are parameters at $\theta = 0$, C an integration constant

However, most of the uniaxial structural viscoelastic models have been based on the discrete element combination approach. Buchthal and Kaiser described the continuous relaxation spectrum of soft tissues by the combination of an infinite number of Voigt and Maxwell elements. The corresponding hysteresis diagram shows an infinite number of bell-shaped curves, which add up to a continuous curve of nearly constant height over a very wide range of frequencies (Fig. 4.10) (Buchthal 51). Sedlin and Sonnerup described the rheological properties of cortical bone with a combination of one linear dashpot η and two linear springs E_1, E_2 (Fig 4.11) (Sedlin 66), leading to (Hirsch 68):

$$\dot{T} + a\,T = b\,\dot{\varepsilon} + c\,\varepsilon \quad \text{with} \quad a = \frac{E_1 + E_2}{\eta} \qquad b = E_2 \qquad c = \frac{E_1\,E_2}{\eta} \tag{4.49}$$

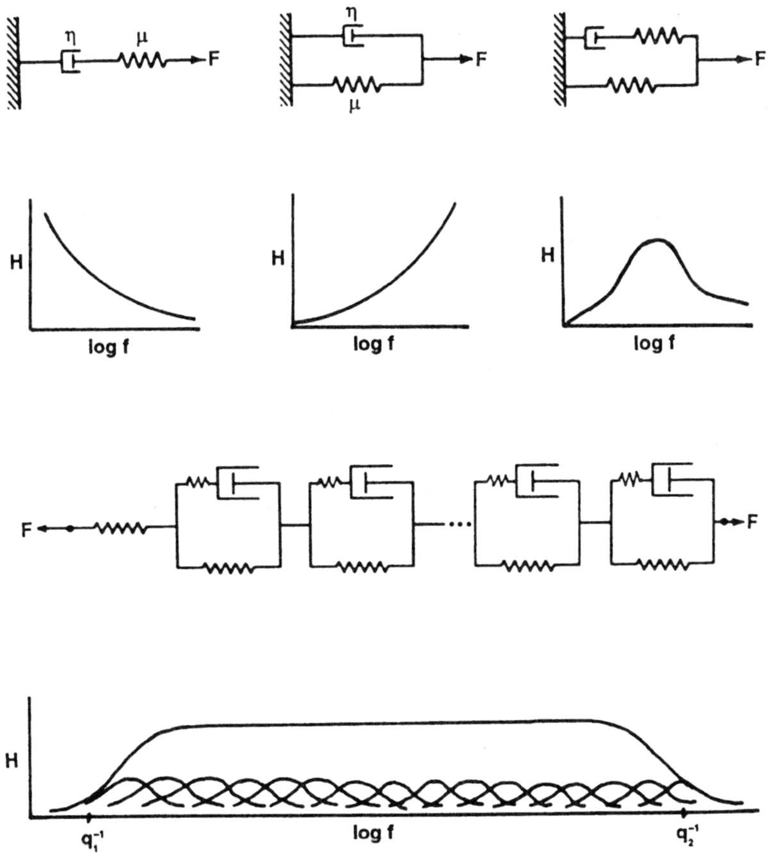

Fig. 4.10. Model for continuous relaxation spectrum
(reprinted from (Fung 93) with permission of Springer-Verlag)

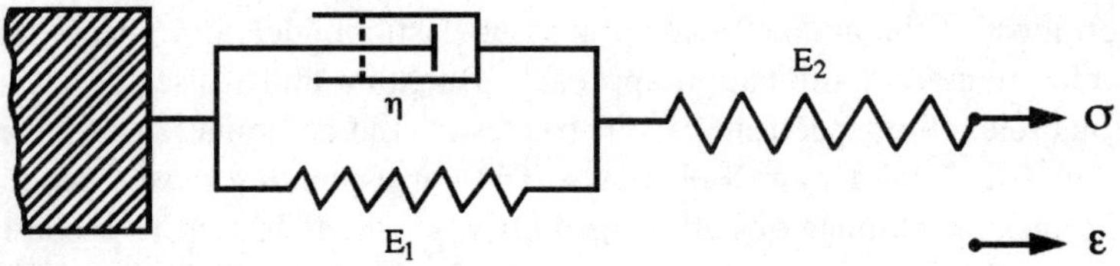

Fig. 4.11. A rheological model for cortical bone (Sedlin 66)
(reprinted from (Hirsch 68) with permission of Elsevier Science)

Jamison et al. proposed a similar model composed of four elements to model the creep behavior of guinea pig skin. The model combines one Maxwell (E_1, η_1) element and one Voigt (E_2, η_2) element in series such as (Jamison 68):

$$X(t) = X_E + X_U + X_R = \frac{C}{E_1}\left(1 + \frac{E_1}{\eta_1} t\right) + \frac{C}{E_2}\left(1 - e^{-\frac{E_2}{\eta_2} t}\right) \tag{4.50}$$

where X is the extension response, X_E is the elastic deformation of E_1
X_U is the recoverable deformation of the Voigt element (E_2, η_2)
X_R is the unrecoverable viscous deformation of η_1

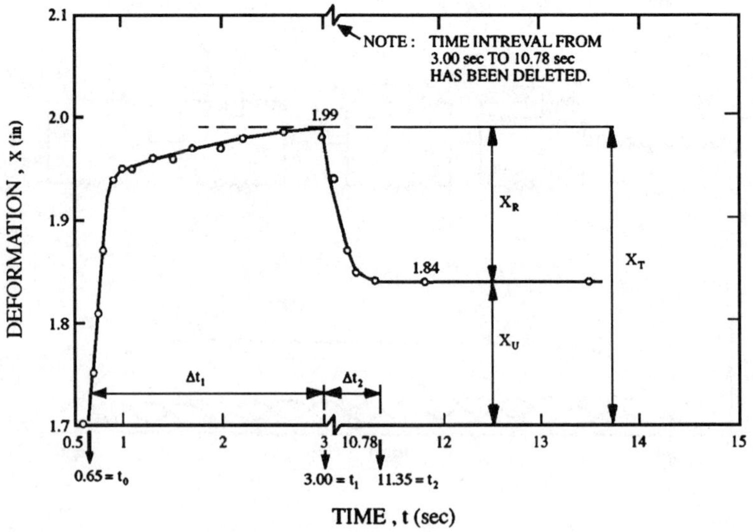

Fig. 4.12. Skin typical creep response
(reprinted from (Jamison 68) with permission of Elsevier Science)

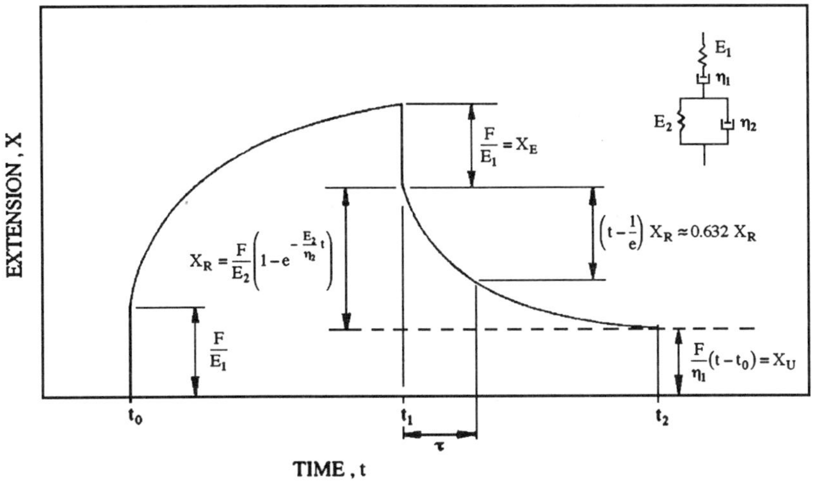

Fig. 4.13. Creep response of a four-element model
(reprinted from (Jamison 68) with permission of Elsevier Science)

By comparing the creep response of the model (Fig. 4.13) with the typical creep curve of skin (Fig. 4.12), it was observed that the skin response does not exhibit instantaneous elasticity. Therefore, the model has been reduced to three elements (Fig. 4.14), which is equivalent to the four-element model when $E_1 \rightarrow \infty$. E_2, η_1 and η_2 have been determined by uniaxial creep and constant rate tests on guinea pig skin specimens taken parallel or perpendicular to the spine (Jamison 68).

The skin creep response has also been described by Sanders who combined elasticity, viscosity and plasticity theories to model the remaining deflection of skin after loading. The total strain was expressed as:

$$\varepsilon = \varepsilon_E + \varepsilon_V + \varepsilon_P \qquad \text{with} \qquad (4.51)$$

$$\varepsilon_V(t) = A_V\left(1 - e^{-\frac{t}{\tau}}\right) \qquad \varepsilon_P(t) = A_P t \qquad (4.52)$$

where ε_E is the initial elastic strain component
ε_V is the viscous strain component with τ as time constant
ε_P is the linearly time-dependent plastic strain component

The model has been applied to torsion creep experiments on *in vivo* human skin (Sanders 73).

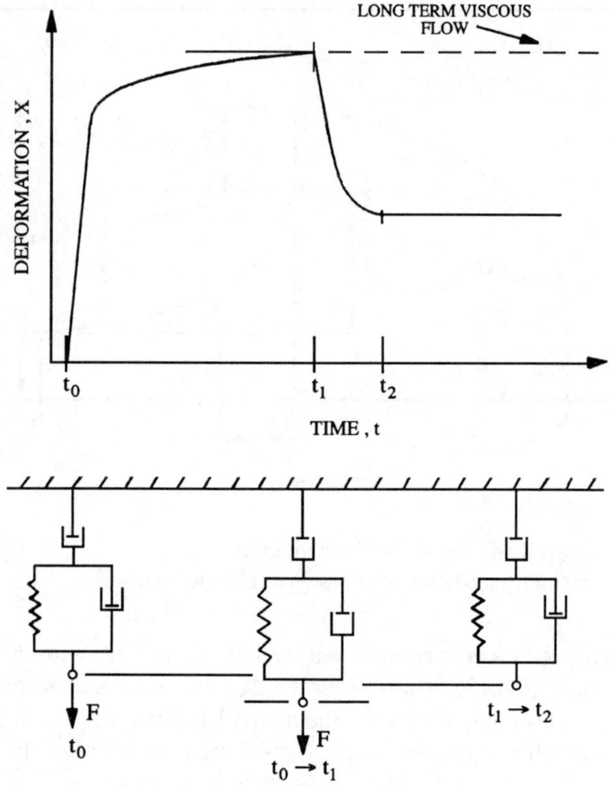

Fig. 4.14. Creep response of a three-element viscous model
(reprinted from (Jamison 68) with permission of Elsevier Science)

Galford and McElhaney used a Maxwell–Kelvin four-element model identical to Jamison's model (Fig. 4.15) and developed a constitutive relationship in the form:

$$T + p_1 \frac{dT}{dt} + p_2 \frac{d^2T}{dt^2} = q_1 \dot{\varepsilon} + q_2 \ddot{\varepsilon} \qquad \text{with} \qquad (4.53)$$

$$\eta_3 = \frac{q_1 q_2}{p_1 q_1 - q_2 - \dfrac{q_1^2 p_2}{q_2}} \qquad \eta_2 = q_1 \qquad E_3 = \frac{q_1}{p_1 - \dfrac{q_2}{q_1} - \dfrac{q_1 p_2}{q_2}} \qquad E_1 = \frac{q_2}{p_2}$$

It has been applied to creep and stress relaxation experiments, among other tissues, on brain from both human and monkey (Galford 70).

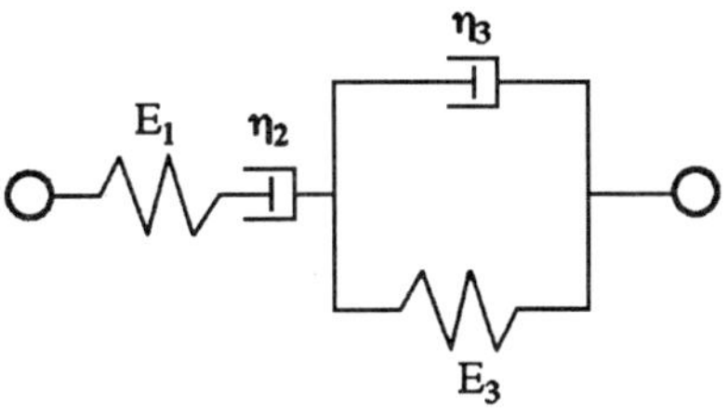

Fig. 4.15. Four-element model of soft tissue
(reprinted from (Galford 70))

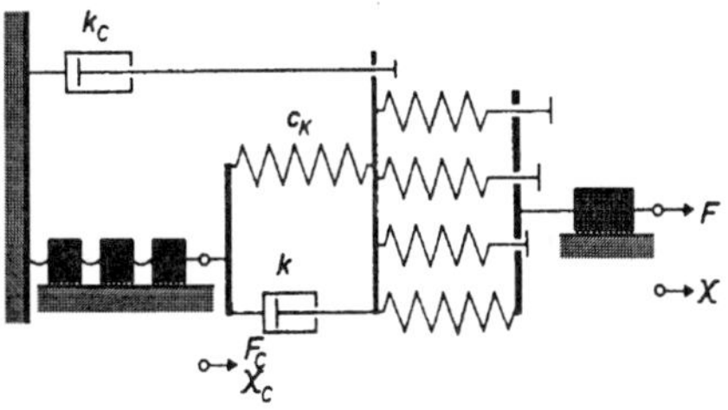

Fig. 4.16. Viidik's model for soft tissues
(reprinted from (Viidik 68))
(with permission of Elsevier Science)

Viidik assumed the parallel fibered soft tissues as composed of individual linear elastic fibers of different resting lengths, successively becoming taut as elongation processes, producing thus their typical non-linear elastic behavior (Fig. 4.17). He proposed an idealized model (Fig. 4.16) composed of Hooke (Fig. 4.18), Newton (Fig. 4.19), Coulomb (Fig. 4.20), and Maxwell (Fig. 4.21) elements to fully describe the non-linear viscoelastic behavior of soft tissues (Viidik 68). Frisen et al. achieved the mathematical analysis of this model, simplified it, and described it with the following differential equation:

$$\frac{dF}{dt}+\frac{F}{\eta_K}(\phi+c_K)=\phi\frac{dX}{dt}+c_K\frac{\phi}{\eta_K}X-c_K\frac{\psi}{\eta_K} \tag{4.54}$$

where F is the load and X is the extension with respect to the resting length. Given some assumptions, this equation has been solved for specific conditions (stress-relaxation, creep) and applied to uniaxial tensile experiments on rabbit anterior cruciate ligaments (Frisen 69a)(Frisen 69b).

Skalak and Chien used a three-element model composed of two linear springs k_1, k_2 and one linear dashpot η to model the incompressible viscoelastic behavior of cytoplasm (Fig. 4.22) (Skalak 82). The behavior was described by:

$$T+\frac{\eta}{k_1}\frac{dT}{dt}=k_1\varepsilon+\eta\left(1+\frac{k_1}{k_2}\right)\frac{d\varepsilon}{dt} \tag{4.55}$$

To describe passive muscle properties, Glantz used a similar model, but with exponential springs instead of linear springs (Fig. 4.23). The behavior was described with the differential equation:

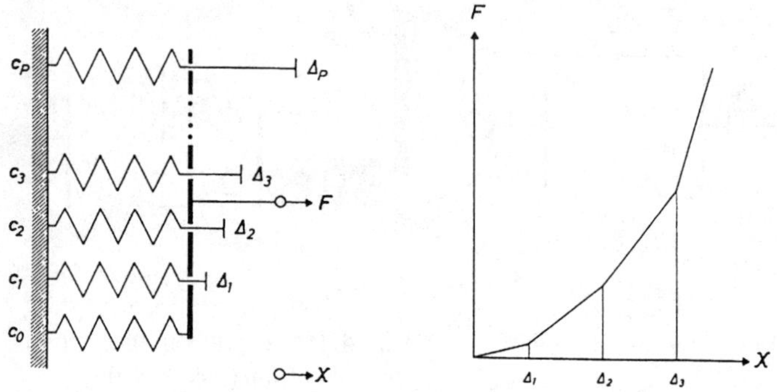

Fig. 4.17. Viidik's non-linear elastic tissue model
(reprinted from (Frisen 69) with permission of Elsevier Science)

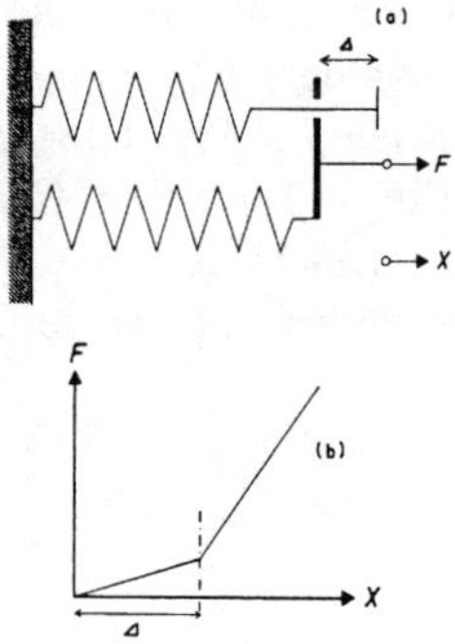

Fig. 4.18. Hooke element

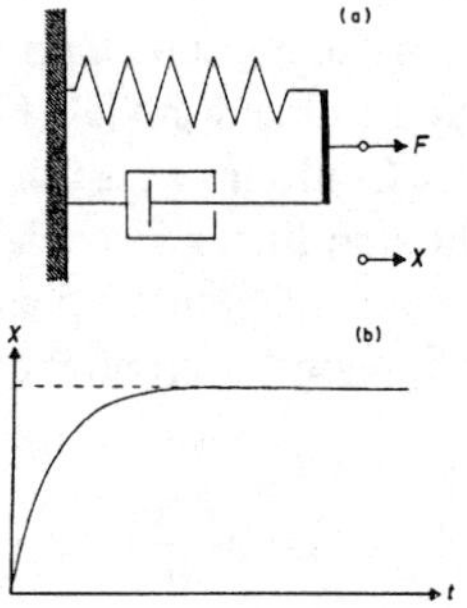

Fig. 4.19. Newton element

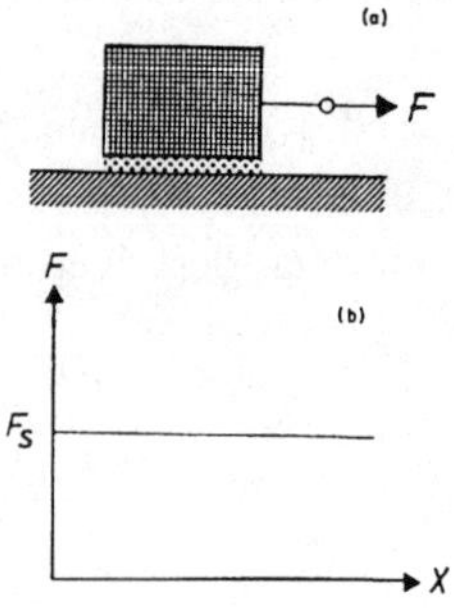

Fig. 4.20. Coulomb element

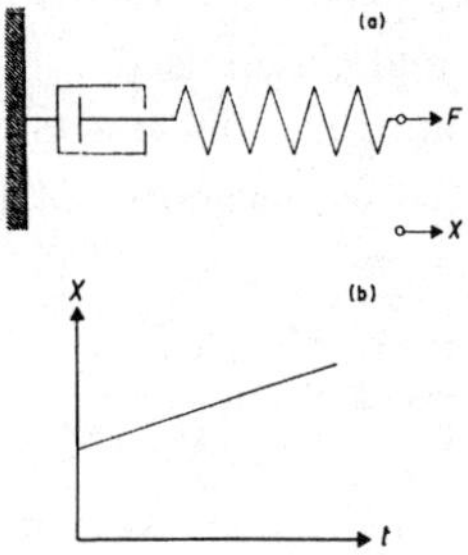

Fig. 4.21. Maxwell element

(Figs. 4.18–4.21 reprinted from (Viidik 68) with permission of Elsevier Science)

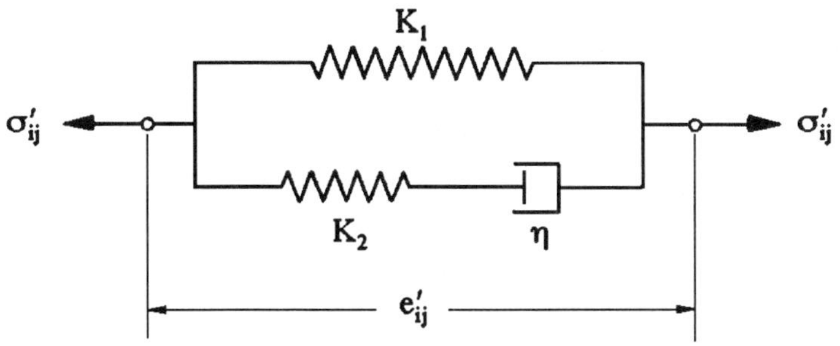

Fig. 4.22. Skalak and Chien's cytoplasm model
(reprinted from (Skalak 82) with permission of Pergamon Press)

$$\frac{dF}{dt} + \frac{\beta}{\gamma}(F+\alpha)^2 = \frac{\beta\,\alpha^2}{\gamma}e^{\beta X} + \beta\,(F+\alpha)\frac{dX}{dt} \tag{4.56}$$

where F is the load and X is the extension with respect to some resting length. This model has been used to fit the stress-relaxation curves of isolated cat papillary muscle and rabbit taeniae coli smooth muscle (Glantz 74) and to analyze the extension of Hill's three-element muscle model (see §1.2) with non-linear viscoelastic passive elements (Glantz 77). Barbenel et al. generalized the discrete element composition to incorporate a logarithmic relaxation spectrum (Barbenel 73). Bingham and De Hoff modeled canine anterior cruciate ligaments as isotropic viscoelastic media with fading memory (Bingham 79).

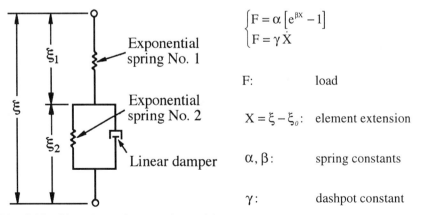

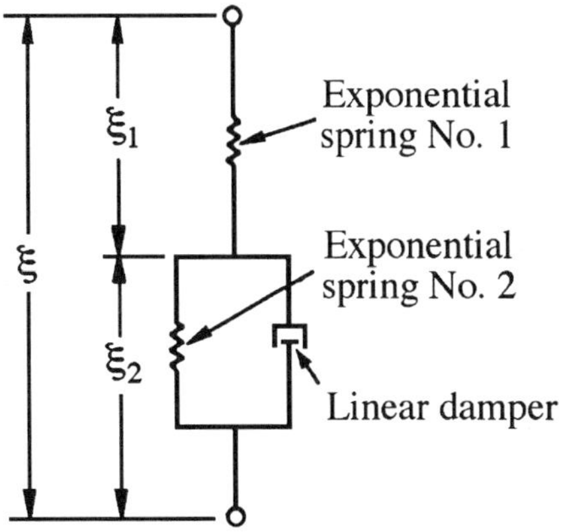

$$\begin{cases} F = \alpha\left[e^{\beta X} - 1\right] \\ F = \gamma\dot{X} \end{cases}$$

F:	load
$X = \xi - \xi_0$:	element extension
α, β:	spring constants
γ:	dashpot constant

Fig. 4.23. Glantz's passive muscle model
(reprinted from (Glantz 74) with permission of Elsevier Science)

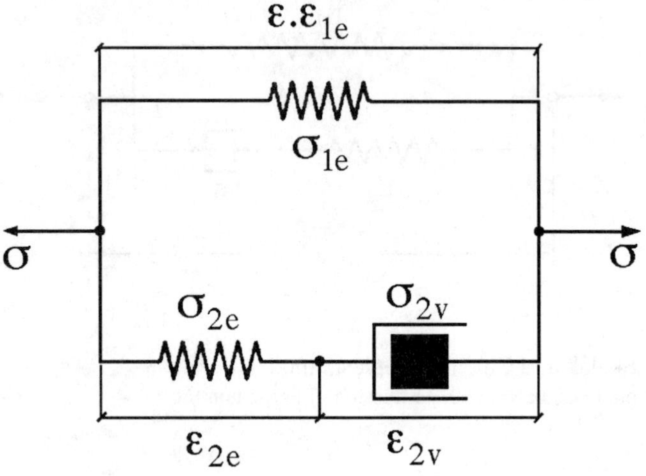

Fig. 4.24. Capelo's passive muscle model
(reprinted from (Capelo 81) with permission of Elsevier Science)

Capelo et al. have also used two non-linear exponential springs, but have chosen a hyperbolic sinusoidal dashpot to represent structural viscosity. The elements were combined as shown in Fig. 4.24 and described by:

$$T^e = E\left[e^{c\varepsilon_e} - 1\right] \quad \text{with distinct coefficients for both springs} \qquad (4.57)$$

$$T^v = \frac{1}{\eta_0} \operatorname{argsh}\left[\eta_0\left(\eta_1 \dot{\varepsilon}_v + \eta_2 \dot{\varepsilon}_v^2\right)\right] \quad \text{for the dashpot} \qquad (4.58)$$

The model has been applied to tensile experiments on rat papillary muscles (Capelo 81).

Sanjeevi et al. considered the global stress in soft tissues as composed of an elastic stress component and a viscous stress component, and have obtained the resulting viscoelastic stress-strain relation as:

$$T = E_1\varepsilon + E_2\varepsilon^2 + \eta_1 \frac{d\varepsilon}{dt} + \eta_2 \varepsilon \frac{d\varepsilon}{dt} \qquad (4.59)$$

It has been used to model the stress-strain response of collagen fibers of different lengths from rat tail tendons (Sanjeevi 82).

4.3 Multi-Dimensional Models

4.3.1 Hyperelastic Modeling

Stress-strain relationships reveal, more than constitute, the mechanical properties of materials: they are only material responses to given mechanical constraints. Hence comes on the idea: to assume soft-tissues as hyperelastic materials and to model them using a strain energy formulation, from which the 3D stress-strain relationships may be derived, as defined in Sect. 2.2.2. The general form of the strain energy function W for an isotropic material has been given in terms of the invariants (2.72) and the extension ratios (2.73) of the strain tensor $\mathbf{C}$ (Ogden 84):

$$\begin{cases} I_C = \lambda_1^2 + \lambda_2^2 + \lambda_3^2 & \equiv I_1 \\ II_C = \lambda_1^2\,\lambda_2^2 + \lambda_1^2\,\lambda_3^2 + \lambda_2^2\,\lambda_3^2 & \equiv I_2 \\ III_C = \lambda_1^2\,\lambda_2^2\,\lambda_3^2 & \equiv I_3 \end{cases} \tag{4.60}$$

$$W(I_1, I_2, I_3) = \sum_{p,q,r=0}^{\infty} c_{pqr}\,(I_1 - 3)^p\,(I_2 - 3)^q\,(I_3 - 1)^r \qquad \begin{cases} c_{000}, a_{000} = 0 \\ (p, q, r) \text{ integers} \end{cases} \tag{4.61}$$

$$W = \sum_{p,q,r=0}^{\infty} a_{pqr} \left\{ \left[\lambda_1^p\,(\lambda_2^q + \lambda_3^q) + \lambda_2^p\,(\lambda_3^q + \lambda_1^q) + \lambda_3^p\,(\lambda_2^q + \lambda_1^q) \right] (\lambda_1\lambda_2\lambda_3)^r - 6 \right\} \tag{4.62}$$

and has been decomposed into distortional W_1 and dilational W_2 components as:

$$W(I_1, I_2, I_3) = W_1(I_1, I_2) + W_2(I_3) \qquad \text{with} \tag{4.63}$$

$$W_1(I_1, I_2) = \sum_{p,q=0}^{\infty} c_{pq0}\,(I_1 - 3)^p\,(I_2 - 3)^q \qquad W_2(I_3) = \sum_{r=1}^{\infty} c_{00r}\,(I_3 - 1)^r \quad \text{i.e.:} \tag{4.64}$$

$$W_1 = \sum_{p=1}^{\infty} 2a_{p00}\left(\lambda_1^p + \lambda_2^p + \lambda_3^p - 3\right) \qquad W_2 = \sum_{r=1}^{\infty} 6\,a_{00r}\left[(\lambda_1\lambda_2\lambda_3)^r - 1\right] \tag{4.65}$$

4.3.2 Phenomenological Elastic Models

Polynomial Form. Various strain energy functions have been formulated on the basis of the polynomial form, namely to model incompressible $(I_3 = 1)$ polymers:

$$W_1 = \frac{1}{4} G \sum_{i=1}^{3} \left(\lambda_i - \frac{1}{\lambda_i} \right)^2 + \frac{1}{4} H \sum_{i=1}^{3} \left(\lambda_i^2 - \frac{1}{\lambda_i^2} \right) \quad \text{i.e., for example:} \quad (4.66)$$

$$W_1 = C_1(I_1 - 3) + C_2(I_2 - 3) \quad \text{Mooney–Rivlin material} \quad \text{(Mooney 40)} \quad (4.67)$$

$$W_1 = C_1(I_1 - 3) \quad \text{neo-Hookean material} \quad \text{(Treloar 75)} \quad (4.68)$$

$$W_1 = C_1(I_2 - 3) + C_2(I_1 - 3) + C_3(I_1 - 3)^2 + C_4(I_1 - 3)^3 \quad \text{(Biderman 58)} \quad (4.69)$$

$$W_1 = C_1(I_1 - 3) + C_2(I_2 - 3) + C_3(I_2 - 3)^2 + C_4(I_2 - 3)^3 \quad \text{(Klosner 69)} \quad (4.70)$$

as well as to model soft tissues:

$$W = \sum_{i=1}^{3} C \left(\lambda_i^\alpha - 1 \right) \quad \text{(Blatz 69)} \quad (4.71)$$

$$W = \frac{E}{n} \sum_{i=1}^{3} \left(\lambda_i^n - 1 \right) \quad \text{for in vivo human skin} \quad \text{(Peng 78)} \quad (4.72)$$

$$\begin{cases} W = V_1 + V_2 + V_3 + V_4 \\ V_1 = A_0 I_1 \\ V_2 = A_1 I_1^2 + A_2 I_2 \\ V_3 = A_3 I_1^3 + A_4 I_1 I_2 + A_5 I_3 \\ V_4 = A_6 I_1^4 + A_7 I_1^2 I_2 + A_8 I_1 I_3 + A_9 I_2^2 \end{cases} \quad \text{for an alveolar finite element model of human lung parenchyma} \quad \text{(Lee 75)} \quad (4.73)$$

$$\begin{cases} W = C_1(I_1 - 3) + C_2(I_2 - 3) + g(I_3) \\ g(I_3) = C_3(I_3 - 1)^2 - (C_1 + 2C_2)(I_3 - 1) \end{cases} \quad \text{for} \begin{cases} \text{vitro cat skin} \\ \text{forearm skin} \end{cases} \quad \text{(Allaire 77)} \quad (4.74)$$

$$W_1 = \sum_{k=1}^{k \le 4} b_k (I_1 - 3)^k \quad \text{for biaxial tests on dog aorta} \quad \text{(Vito 80)} \quad (4.75)$$

$$W = \frac{\mu}{k} \left(\lambda_1^k + \lambda_2^k + \lambda_3^k - 3 \right) \quad \text{for canine left ventricle} \quad \text{(Needleman 83)} \quad (4.76)$$

Exponential Forms. Snyder presented how Fung's uniaxial exponential relation defined in (4.13) and (4.14) could be integrated in order to obtain a strain energy function in terms of the extension ratio λ such as:

$$W_1 = \frac{B}{A^2}\left[e^{A(\lambda-1)} - A\,\lambda\right] \qquad T = \frac{B}{A}\left[e^{A(\lambda-1)} - 1\right] \qquad \text{(Snyder 72)} \qquad (4.77)$$

It was then extended to three dimensions by Snyder and Lee with the form:

$$W_1 = \frac{B}{A^2}\left[e^{a(\Gamma-1)} - A\,\Gamma\right] \qquad \Gamma = \frac{I_1 + \sqrt{I_1^2 - 3I_2}}{I_2} \qquad \text{(Snyder 75)} \qquad (4.78)$$

for which one $\Gamma = \lambda$ in uniaxial extension. It has been applied to tensile experiments on frog skin (Snyder 75). Various other similar exponential forms have been applied to soft tissue modeling, such as:

$$W_1 = \sum_{i=1}^{3} C\left(e^{\alpha\left(\lambda_i^2-1\right)} - 1\right) \qquad \text{(Blatz 69)} \qquad (4.79)$$

$$W_1 = C\left[e^{KI_1(I_1^2-3I_2)} - 1\right] \qquad \text{(Gou 70)} \qquad (4.80)$$

$$\begin{cases} W = C_1\left[e^{\beta(I_1-3)} - 1\right] + C_2(I_2-3) + g(I_3) \\ g(1) = 0 \end{cases} \qquad \text{for cat skin} \quad \text{(Veronda 70)} \qquad (4.81)$$

$$\begin{cases} W_1 = \dfrac{\beta}{2\alpha}\left(e^{\alpha(I_1-3)} - 1\right) \\ W_1 = \dfrac{\beta}{2\alpha}e^{\alpha(I_1-3)} \end{cases} \qquad (4.82)$$

for dog left ventricle (Demiray 72), bovine heart valves (Demiray 76), porcine pulmonary valves (Christie 82)

$$W = \alpha\left(e^{\beta[I_1-3]+\gamma[I_2-3]} - 1\right) \qquad \text{for tensile experiments on arteries (Vito 73)} \qquad (4.83)$$

$$W = \frac{E}{\nu\beta^2}\left[\frac{\nu}{(1+\nu)}\sum_{i=1}^{3} e^{\beta E_i} + \frac{(1-2\nu)}{(1+\nu)}e^{\left(-\frac{\beta\nu}{1-2\nu}\right)\sum_{i=1}^{3} E_i} - 1\right] \qquad (4.84)$$

cylindrical description for rat left ventricle (Janz 74)

$$W_1 = \frac{1}{2}\left(\lambda\, I_1^2 + 2\mu\, I_2\right)\left[\beta + e^{\alpha I_1^2 + \gamma I_2}\right] \qquad \text{for lung parenchyma} \quad \text{(Karakaplan 80)} \ (4.85)$$

$$W_1 = \frac{C}{2D_A}e^{\left(\alpha I_1^2 + \beta I_2\right)} \qquad \text{for lung parenchyma} \quad \text{(Vawter 80)} \qquad (4.86)$$

$$W_1 = \frac{C}{\alpha} e^{\alpha\left(\sqrt{I_3} + \frac{nI_2}{2I_3} + \frac{(n-1)I_1}{2} - 3n + \frac{1}{2}\right)} \quad \text{for rabbit papillary muscle} \qquad \text{(Yang 91)} \tag{4.87}$$

Tong and Fung generalized the form of the strain energy function for soft tissues in terms of the Green–Lagrange strain components E_{ij} as (Tong 76):

$$W = \frac{1}{2} \sum_{i,j,k,l=1}^{3} \alpha_{ijkl} E_{ij} E_{kl} + \beta_0 \sum_{m,n,p,q=1}^{3} \beta_{mnpq} E_{mn} E_{pq} e^{\left(\sum_{i,j=1}^{3} \nu_{ij} E_{ij} + \frac{1}{2} \sum_{i,j,l,k=1}^{3} \gamma_{ijkl} E_{ij} E_{kl} + \ldots\right)} \tag{4.88}$$

where α_{ijkl}, β_{mnpq}, ν_{ij}, γ_{ijkl} and β_0 are material constants. It has been applied to biaxial experiments on skin in the form:

$$W = f(\alpha, \mathbf{E}) + \frac{c}{2} e^{F(a,\mathbf{E})} \qquad f(\alpha, \mathbf{E}) = \frac{1}{2}\left(\alpha_1 E_{11}^2 + \alpha_2 E_{22}^2 + 2\alpha_4 E_{11} E_{22}\right) \tag{4.89}$$

$$F(a, \gamma, \mathbf{E}) = a_1 E_{11}^2 + a_2 E_{22}^2 + a_3 E_{12}^2 + 2a_4 E_{11} E_{22} + \gamma_1 E_{11}^3 + \gamma_2 E_{22}^3 + \gamma_4 E_{11}^2 E_{22} + \gamma_5 E_{11} E_{22}^2$$

where α_i, a_i, γ_i and c are experimental constants. The second term was observed to express the behavior of the material at a high stress level, while the first one has been introduced to account for the "biphasic" aspect of the data at a lower stress. In practice, the third-degree terms are assumed negligible with no significant loss of accuracy. This leads therefore to the simplified strain energy:

$$W \approx f(\alpha, \mathbf{E}) + \frac{c}{2} e^{F(a,\mathbf{E})} \qquad \begin{cases} f(\alpha, \mathbf{E}) = \frac{1}{2}\left(\alpha_1 E_{11}^2 + \alpha_2 E_{22}^2 + 2\alpha_4 E_{11} E_{22}\right) \\ F(a, \mathbf{E}) = a_1 E_{11}^2 + a_2 E_{22}^2 + a_3 E_{12}^2 + 2a_4 E_{11} E_{22} \end{cases} \tag{4.90}$$

Constants have been determined from biaxial tensile tests on rabbit abdominal skin, assumed orthotropic (Tong 76). Fung's model has also been applied as:

$$\begin{cases} f(\alpha, \mathbf{E}) = 0 & \text{for arteries and veins} \qquad \text{(Chuong 83)} \\ F(a, \mathbf{E}) = a_1 E_1^2 + a_2 E_2^2 + a_3 E_3^2 + 2a_4 E_1 E_2 + 2a_5 E_2 E_3 + 2a_6 E_3 E_1 \\ F(b, \mathbf{E}) = b_1 E_{\theta\theta}^2 + b_2 E_{ZZ}^2 + b_3 E_{RR}^2 + 2b_4 E_{\theta\theta} E_{ZZ} + 2b_5 E_{ZZ} E_{RR} + 2b_6 E_{RR} E_{\theta\theta} \end{cases} \tag{4.91}$$

$$\begin{cases} f(\alpha, \mathbf{E}) = 0 & \text{for a two dimensional finite} \\ F(a, \mathbf{E}) = C_2 I_1^2 + C_3 I_2 + C_4 I_3 & \text{element model of heart valve} \end{cases} \qquad \text{(Huang 90)} \tag{4.92}$$

$$\begin{cases} f(\alpha, \mathbf{E}) = -\frac{c}{2} & \text{for dog ventricle} \\ F(a, \mathbf{E}) = 2a\left(E_{RR} + E_{\theta\theta} + E_{ZZ}\right) & \text{in cylindrical coordinate system} \end{cases} \quad \text{(Guccione 91)} \quad (4.93)$$

Danielsen also modeled skin as non-linear anisotropic, elastic membrane under large deformation with an exponential function as suggested by Fung (Danielsen 73). However, as skin cannot support negative stress and hence buckles easily, many practical problems could not be solved this way. They then proposed to apply a generalized tension field theory to Z-plasty operations on membranes undergoing arbitrarily large deformation, and compared the results with published data (Danielsen 75).

4.3.3 Structural Elastic Models

An interesting multi-dimensional theory has been proposed by Lanir for fibrous connective tissue modeling, on the basis of microstructural and thermodynamic observations (Lanir 83a). Assuming tissues as composed of several networks of different types of fibers embedded in a fluid matrix, he has developed a strain energy function including their structural properties such as their angular and geometrical non-uniformities. For a volume unit of tissue, the distortional energy (2.75) was expressed as the sum of the strain energies of the different fiber types:

$$W_1 = \sum_k W_k \qquad \text{with} \qquad W_k = \int_\Omega S_k R_k(\mathbf{u}) w_k(\lambda)\, d\Omega \qquad (4.94)$$

where
S_k is the volumetric fraction of unstrained k-type fibers
$R_k(\mathbf{u})$ is the density of with unit direction $\mathbf{u}$
$w_k(\lambda)$ is the k-type fibers uniaxial strain energy function
Ω represents the range of fiber orientations
λ is the individual fiber stretch ratio

The model was then extended to account for the non-uniform undulation of each k-type fiber along its direction $\mathbf{u}$ by means of a fiber strain-energy function $w_k^*(\lambda)$ incorporating an undulation density distribution function $D_{k,u}(X)$ (Lanir 83a). Further extensions concerned viscoelastic behavior and non-homogeneities by means of theoretical functionals introduced in the fiber strain energy functions.

Following this approach, Lanir has developed a biaxial incompressible viscoelastic model for skin assumed as a membranous tissue composed of a ground substance and planar collagen and elastin fibers (Lanir 79). Collagen fibers were assumed in slack undulated state and elastin fibers in prestretched state. The compressive and bending rigidity of the fibers were not considered and the matrix was assumed to contribute to stress only through the hydrostatic pressure. The advantage of this structural model lies in the fact that the non-homogenous non-linear anisotropic properties of the skin components may be included while avoiding ambiguity in material characterization and offering an acute insight into their structure, mechanics, and function. A similar formulation has also been applied to model the lung tissue (Lanir 83b).

Horowitz et al. led an interesting application of Lanir's fibrous description to model the anisotropic properties of the myocardial muscle (Horowitz 88c). Here also, the cardiac tissue was assumed to be composed of muscle fibers and collagen fibers embedded in a gelatinous ground substance. Assuming incompressibility, the second Piola–Kirchhoff stress tensor $\mathbf{S}$ was defined in the form (2.85):

$$\mathbf{S} = \frac{\partial W_1}{\partial \mathbf{E}} + L\frac{\partial I_3}{\partial \mathbf{E}} \tag{4.95}$$

where: L is a Lagrange multiplier related to the matrix hydrostatic pressure
I_3 is the 3rd invariant of the Cauchy–Green right dilation tensor $\mathbf{C}$
$\mathbf{E}$ is the global tissue Green–Lagrange strain tensor
W_1 is the fiber distortional energy function from Lanir (4.94)

Assuming a fiber waviness distribution function $D_{k,u}$ and a straight fiber constant stiffness C_k, the wavy fiber's stress-strain relations were expressed in the form:

$$f_k^*(\varepsilon') = \int_0^{\varepsilon'} D_{k,u}(x)\, f_k(\varepsilon)\, dx \qquad \text{where} \tag{4.96}$$

$$f_k(\varepsilon) = C_k \varepsilon \tag{4.97}$$

is the stress-strain relation for straight fibers. Then, considering the wavy fiber stress as deriving from the strain energy $w_k^*(\varepsilon)$:

$$f_k^*(\varepsilon') = \frac{\partial w_k^*(\varepsilon)}{\partial \varepsilon'} \tag{4.98}$$

the global tissue stress tensor could be developed into the following form:

$$\mathbf{S} = \int_\Omega S_k R_k(\mathbf{u})\, f_k^*(\varepsilon') \frac{\partial \varepsilon'}{\partial \mathbf{E}}\, d\Omega + L\frac{\partial I_3}{\partial \mathbf{E}} \tag{4.99}$$

$\varepsilon = \dfrac{\varepsilon' - x}{1 + 2x}$ true wavy fiber strain

$\varepsilon' = \dfrac{\partial \xi^r}{\partial \xi_1'} \dfrac{\partial \xi^s}{\partial \xi_1'} E_{rs}$ total fiber strain

E_{rs} global tissue strain component

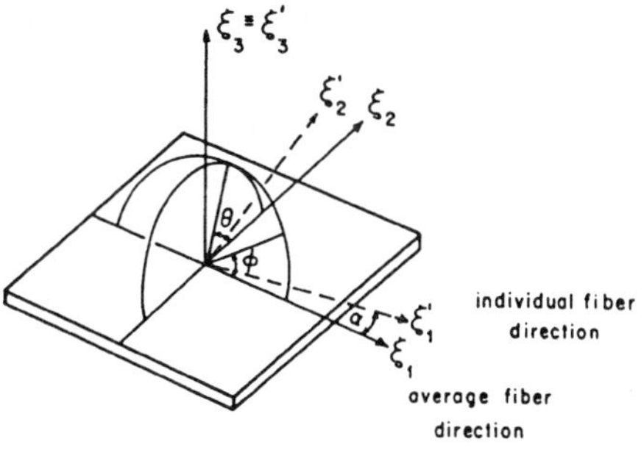

Fig. 4.25. Orientations of the fiber coordinate systems
(reprinted from (Horowitz 88c), with permission of ASME)

The fiber orientation and waviness distribution functions were assumed to be:

$$D_{k,u}(x) = \frac{1}{\sigma_k \sqrt{2\pi}} e^{-\frac{(\mu_k - x)^2}{2\sigma_k^2}} \quad \text{with } \mu_k, \sigma_k, \text{ mean and variance of distribution} \tag{4.100}$$

$$R_1(\alpha) = \frac{12}{\pi} \quad \text{for } -\frac{\pi}{24} \le \alpha \le \frac{\pi}{24} \quad \text{for muscle fibers,} \quad \text{and for collagen} \tag{4.101}$$

$$R_2(\theta', \phi') = R_2(\theta')\, R_2(\phi') \quad \text{for } 0 \le \theta' \le 2\pi \quad \text{and} \quad 0 \le \phi' \le \pi \quad \text{with} \tag{4.102}$$

$$R_2(\phi') = \begin{cases} \dfrac{1}{2\sqrt{2\pi}\sigma_{\phi'}} e^{-\frac{(\mu_{f} - \phi')^2}{2\sigma_{\phi'}^2}} & 0 \le \phi' \le \dfrac{\pi}{2} \\ \dfrac{1}{2\sqrt{2\pi}\sigma_{\phi'}} e^{-\frac{(\pi - \mu_{f} - \phi')^2}{2\sigma_{\phi'}^2}} & \dfrac{\pi}{2} \le \phi' \le \pi \end{cases} \quad \text{and} \quad R_2(\theta') = \frac{1}{2\pi} \quad \text{with} \tag{4.103}$$

$$\theta' = \arctan\left(\frac{\sin\theta \sin\phi}{\cos\alpha \cos\theta \sin\phi - \sin\alpha \cos\phi} \right) \quad \phi' = \arccos(\sin\alpha \cos\theta \sin\phi + \cos\alpha \cos\phi)$$

with $\xi_1(\rho, \theta, \phi)$ the global spherical coordinate systems
$\xi_1'(\rho', \theta', \phi')$ the individual muscle fiber coordinate system
α the angle between the polar axis ξ_1 and ξ_1'
$\mu_{\phi'}$, $\pi - \mu_{\phi'}$ the means of the bimodal collagen fiber distribution
$\sigma_{\phi'}$ the variance of the bimodal collagen fibers distribution

The model has been applied to two- (Horowitz 88a) and three-dimensional (Horowitz 88b) finite element modeling of passive myocardium and ventricle.

Also assuming muscle constituted of muscle fibers embedded in a gelatinous matrix, Humphrey et al. applied an energy function composed of two exponential functions, one of which being a function of the fiber extension ratio λ:

$$W_1 = W_m + W_f \qquad \text{with} \tag{4.104}$$

$$W_m = c\left[e^{b(I_1-3)} - 1\right] \qquad \text{as strain energy contribution of the matrix} \tag{4.105}$$

$$W_f = A\left[e^{a(\lambda-1)^2} - 1\right] \qquad \text{as strain energy contribution of the fibers} \tag{4.106}$$

It has been applied to the identification of various canine cardiac tissues (Humphrey 87). Later, they changed their exponential function to a polynomial form, to include a coupling term between I_1 and λ as:

$$W_1 = c_1(\lambda+1)^2 + c_2(\lambda-1)^3 + c_3(I_1-3) + c_4(I_1-3)(\lambda-1) + c_5(I_1-3)^2 \tag{4.107}$$

It has been applied to passive canine myocardium (Humphrey 90), to sinus and cosinus regions of the right ventricle free wall (Sacks 93), and to various other regions of the myocardium (Novak 94).

4.3.4 Viscoelastic Models

Phenomenological Models. Zeng et al. applied Fung's QLV theory (Sect. 4.1.2) formulated for creep with an elastic response derived from a strain energy formulation. The Lagrangian stress was defined as:

$$T = \int_{-\infty}^{t} J(t-\tau)\frac{dT^{(e)}[E(\tau)]}{d\tau}\,d\tau \qquad \text{with} \tag{4.108}$$

$$J(t) = A - (A-1)\,e^{-b(t-\tau)} \qquad \text{as reduced creep function, and} \tag{4.109}$$

$$\mathbf{T}^{(e)} = \frac{\partial W}{\partial \mathbf{E}}\frac{\partial \mathbf{E}}{\partial \lambda} \qquad \text{as elastic response, derived from} \tag{4.110}$$

$$W = \frac{C}{2}\left[e^{Q_{12}} + e^{Q_{13}} + e^{Q_{32}}\right] \qquad \text{with} \quad Q_{ij} = a_1 E_{ii}^2 + a_2 E_{jj}^2 + 2a_4 E_{jj} E_{ii} \tag{4.111}$$

It was applied to human and dog lung characterization (Zeng 87). Huyghe et al. assumed the stress as composed of an isotropic elastic stress $\mathbf{S}^c(\mathbf{E})$ and a quasi-linear viscoelastic stress $\mathbf{S}^s(\mathbf{E},t)$ as:

$$\mathbf{S} = \mathbf{S}^c(\mathbf{E}) + \mathbf{S}^s(\mathbf{E},t) \qquad \text{with} \tag{4.112}$$

$$\mathbf{S}^c = \frac{\partial W^c}{\partial \mathbf{E}} \qquad W^c = \frac{c^c}{2}(J-1)^2 \qquad J = \det(\mathbf{F}) = \sqrt{I_3} \tag{4.113}$$

$$\mathbf{S}^s(\mathbf{E},t) = \int_{-\infty}^{t} G(t-\tau)\frac{\partial \mathbf{S}^{(e)}[\mathbf{E}(\tau)]}{\partial \tau}\,d\tau \qquad \text{with} \tag{4.114}$$

$$G(t) = \frac{1 + \int_0^\infty S(\tau)e^{-\frac{t}{\tau}}\,d\tau}{1 + \int_0^\infty S(\tau)\,d\tau} \qquad S(t) = \begin{cases} \frac{c}{t} + S_0 & \tau_1 \le t \le \tau_2 \\ 0 & t < \tau_1, t > \tau_2 \end{cases} \tag{4.115}$$

$$\mathbf{S}^{(e)} = \frac{\partial W_1}{\partial \mathbf{E}} \qquad \text{with} \tag{4.116}$$

$$\begin{aligned} W_1 = {} & c^s\left[e^{a^s(E_{12}E_{12}+E_{13}E_{13}+E_{23}E_{23})} - 1\right] + c^n\Big[\left(e^{a^{cf}E_{11}} - a^{cf}E_{11}\right) + \left(e^{a^{cf}E_{22}} - a^{cf}E_{22}\right) + \left(e^{a^{f}E_{33}} - a^{f}E_{33}\right) + \\ & + \left(e^{a^bE_{11}} - a^bE_{11}\right)\left(e^{a^bE_{22}} - a^bE_{22}\right) + \left(e^{a^bE_{11}} - a^bE_{11}\right)\left(e^{a^bE_{33}} - a^bE_{33}\right) + \\ & + \left(e^{a^bE_{22}} - a^bE_{22}\right)\left(e^{a^bE_{33}} - a^bE_{33}\right) - 6\Big] \end{aligned} \tag{4.117}$$

and
- c^c as volumetric modulus
- c^n as initial normal stiffness
- c^s as initial shear stiffness
- a^{cf} as exponential factor in cross-fiber stiffness
- a^f as exponential factor in fiber stiffness
- a^b as exponential factor in biaxial stiffness
- a^s as exponential factor in shear stiffness

This model has been applied to passive heart tissue characterization. The results however were not always satisfying. A doubt concerning the applicability of the QLV theory to passive heart muscle modeling was expressed (Huygue 91).

Johnson et al. developed a single integral finite strain viscoelastic theory, and formulated a general non-linear 3D finite viscoelastic constitutive relation as:

$$\begin{aligned}\boldsymbol{\sigma} = & -p\mathbf{I} + C_0\left[(1+\mu I(t))\,\mathbf{B}(t) - \mu\,\mathbf{B}^2(t)\right] \\ & -(C_0 - C_\infty)\int_0^t G(t)\left[(1+\mu I(\tau))\,\mathbf{B}(\tau) - \mu\,\mathbf{B}^2(\tau)\right] d\tau\end{aligned} \tag{4.118}$$

where $\mathbf{I}$ is the identity tensor, $\mathbf{B}$ the Cauchy–Green left dilation tensor,
$G(t)$ the relaxation function, μ the shear modulus,
C_0 the instantaneous modulus, C_∞ the long time modulus,
p the hydrostatic pressure due to incompressibility,
$I(\tau) = \mathrm{Tr}(\mathbf{C})$ the third invariant of the right dilation tensor $\mathbf{C}$.

This relation has been applied to human patellar tendon modeling (Johnson 92). Another single integral finite strain constitutive form has also been developed by Pioletti who proposed to distinguish the short term and long term memory effects:

$$\mathbf{S} = \mathbf{S}^{(e)} + \mathbf{S}^{(v)} + \mathbf{S}^{(r)} \quad \text{with} \tag{4.119}$$

$$\mathbf{S}^{(e)}(\mathbf{C}) = -p\mathbf{C}^{-1} + \alpha\beta\left(2e^{[\beta(I_1-3)]} - I_1\right)\mathbf{I} + \alpha\beta\,\mathbf{C} \qquad \mathbf{S}^{(v)}(\mathbf{C},\dot{\mathbf{C}}) = \eta\,(I_1 - 3)\,\dot{\mathbf{C}}$$

$$\mathbf{S}^{(r)}(\mathbf{C}) = \int_0^t \dot{G}(\tau)\mathbf{S}^{(e)}[\mathbf{C}(t-\tau)]\,d\tau \quad \text{with} \quad G(t) = \frac{\sum_{k=1}^{3} a_k e^{-\frac{t}{\tau_k}}}{\sum_{k=1}^{3} a_k}$$

where $\mathbf{S}^{(e)}$ is the elastic response in terms of the dilation tensor $\mathbf{C}$
$\mathbf{S}^{(v)}$ is the effects of the strain rate $\dot{\mathbf{C}}$ (short term memory effects)
$\mathbf{S}^{(r)}$ is the effect of stress relaxation (long term memory effects)
$G(t)$ is the reduced relaxation function as a series of exponentials
p is the hydrostatic pressure, α and β are parameters of elasticity
η is parameter of viscosity, τ_k are relaxation moduli, a_k are constants

This model satisfies the basic mechanical and thermodynamical requirements and is meaningful in large strains situations. Its originality is based on the fact that the different mechanical behaviors (immediate, short term memory, and long term memory contributions) are described in one framework allowing a compact description of the biomechanical properties of different soft tissues. Identification of the parameters has been performed for traction and relaxation experiments on tendons and ligaments (Pioletti 97).

Structural Model. Shoemaker et al. assumed the stress composed of a compliant component due to the gelatinous matrix and a fibrous component due to the fibers. The Piola–Kirchhoff second stress tensor $\mathbf{S}$ was then expressed as:

$$\mathbf{S} = \mathbf{S}^F + \mathbf{S}^C \quad \text{with} \tag{4.120}$$

$$\mathbf{S}^F = \int_{-\frac{\pi}{2}}^{\frac{\pi}{2}} D(\theta) S_f \frac{\partial E_f}{\partial \mathbf{E}} \, d\theta \quad \text{fibrous stress similar to Lanir's form (§4.33)} \tag{4.121}$$

$$\mathbf{S}^C = \mathbf{\Lambda} \int_{-\infty}^{t} g(t-\tau) \frac{\partial \mathbf{E}(\tau)}{\partial \tau} d\tau \quad \text{compliant stress similar to Fung's QLV form} \tag{4.122}$$

where: $D(\theta)$ is the fiber orientation distribution function
θ the angle of a fiber with respect to the x axis
S_f the fiber stress measure
$g(t)$ the reduced relaxation function
E_f the fiber strain measure
$\mathbf{\Lambda}$ a constant tensor

The fiber stress measure S_f was also assumed to be a viscoelastic function of E_f:

$$S_f = \int_{-\infty}^{t} G(t-\tau) \frac{\partial E_f(\tau)}{\partial \tau} d\tau \quad \text{with} \quad G(t) = G_0 g(t) \quad \text{and} \tag{4.123}$$

$$E_f = \frac{1}{\lambda_0^2} \left[(1-b)\,\Sigma + b\,\Sigma^- - \Sigma_0 \right] \quad \text{for} \quad S_f > 0 \quad \text{with} \tag{4.124}$$

$$\Sigma = \frac{1}{2}\left(\lambda^2 - 1\right) \quad \text{and} \quad \lambda_0 = \sqrt{2\Sigma_0 + 1} \tag{4.125}$$

where Σ and λ are the fiber strain and stretch ratio
Σ_0 and λ_0 the effective straightening fiber strain and stretch ratio
Σ^- the perpendicular line strain, G_0 and b constants $(0 \le b \le 1)$

The reduced relaxation function $g(t)$ was taken in a form similar to Fung's QLV:

$$g(t) = \frac{1 + a\int_0^{\infty} f(\tau) e^{-\frac{t}{\tau}} \, d\tau}{1 + a\int_0^{\infty} f(\tau) \, d\tau} \quad \text{with} \quad f(t) = e^{-pt} \quad \text{and a} \quad \text{a constant} \tag{4.126}$$

Finally, assuming:

$$\Sigma = E_{11}\cos^2\theta + E_{22}\sin^2\theta + (E_{12}+E_{21})\sin\theta\cos\theta \tag{4.127}$$

$$\Sigma^- = E_{11}\sin^2\theta + E_{22}\cos^2\theta - (E_{12}+E_{21})\sin\theta\cos\theta \tag{4.128}$$

$$E_0 = \begin{cases} E_1(\cos\theta + E_2\sin\theta)^{-1} & \text{for } \; 0 \le \theta \le \pi/2 \\ E_1(\cos\theta - E_2\sin\theta)^{-1} & \text{for } \; -\pi/2 < \theta < 0 \end{cases} \tag{4.129}$$

$$D(\theta) = \frac{c}{\pi} \qquad k = G_0\frac{c}{\pi} \qquad \text{with } c,\ E_1, \text{ and } E_2 \text{ material constants} \tag{4.130}$$

biaxial experiments have been done on human skin and canine pericardium (Shoemaker 86).

Conclusion

Considering the numerous approaches which have been followed towards soft tissue modeling, as well as the specific nature and conditions of the experiments, it is probably not possible to present one model as more reliable for any case than the others. It is also not possible to collect and review all the existing biomechanical models without losing the understanding of the natural mechanical behavior common to all of them. The purpose of this chapter is therefore not to list all the available models, but rather to outline more or less the different forms they can take. The first distinction may be done between uniaxial / multi-dimensional models. Then, models may be distinguished on whether they are phenomenological / structural or elastic / viscoelastic, and with respect to the tissue they model. Most of the approaches followed could have been applied indiscriminately to tendon, muscle, heart, skin, lung... However, due to its activable contractile properties, muscle requires a specific analysis, in order to characterize the active behavior separately from the passive response. This is the purpose of the next chapter.

References

This chapter is essentially based on (Crisp 72), (Fung 93), (Gallagher 82), (Silver 87), (Viidik 80, 87), (Woo 93), (CHARM D4).

Allaire 77 P.E. Allaire, J.G. Thacker, R.F. Edlich, G.J. Rodenheaver, M.T. Edgerton (1977), Finite deformation theory for in-vivo human skin, *J. Bioeng.*, 1, 239–249

Barbenel 73 J.C. Barbenel, J.H. Evans, J.B. Finlay (1973), Stress-strain time relations for soft connective tissues, in *Perspectives in Biomedical Engineering,* ed. by R.M. Kenedi. London: Macmillan

Beskos 75 D.E Beskos, J.T. Jenkins (1975), A mechanical model for mammalian tendon, *J. Appl. Mech.,* 42, 755–758

Best 94 T.M. Best, J. McElhaney, W.E. Garrett Jr, B.S. Myers (1994), Characterization of the passive responses of live skeletal muscle using the quasilinear theory of viscoelasticity, *J. Biomechanics,* 27, 413–419

Biderman 58 V.L. Biderman (1958), *Calculation of Rubber Parts,* in Russian: *Rascheti na Prochnos*t, Moscow

Bingham 79 D.N. Bingham, P.H. De Hoff (1979), A constitutive equation for the canine anterior cruciate ligament, *J. Biomech. Engng.*, 101, 15–22

Blatz 69 P.J. Blatz, B.M. Chu, H. Wayland (1969), On the mechanical behavior of elastic animal tissue, *Trans. Soc. Rheol.,* 13, 82–102

Buchthal 51 F. Buchthal, E. Kaiser (1951), The rheology of the cross-striated muscle fibre with particular reference to isotonic conditions, *Dan. Biol. Med.*, 21, 7, 233–291

Capelo 81 A. Capelo, V. Comincioli, R. Minelli, C. Poggesi, C. Reggiani, L. Ricciardi (1981), Study and parameters identification of a rheological model for excised quiescent cardiac muscle, *J. Biomechanics*, 14, 1–11

Carton 62 Carton (1962), Elastic properties of single elastic fibers, *J. Applied Physiol.,* 17, 547–551

Chadwick 81 R.S. Chadwick (1981), The myocardium as a fluid-fiber continuum: passive equilibrium configurations, in *Advances in Bioengineering*, ed. by D.C. Viano. New York: ASME

CHARM D4 W. Maurel, Y. Wu (1994), Survey of mechanical models for tissue deformation and muscle contraction with identification of parametric data, LIG-EPFL/MIRALab-UG, ESPRIT 9036 Project CHARM, Deliverable D4

Christie 82 G.W. Christie, I.C. Medland (1982), A non-linear finite element stress analysis of bioprosthetic heart valves, in *Finite Elements in Biomechanics*, ed. by R.H. Gallagher, B.R. Simon, P.C. Johnson, J.F. Gross. Chichester, UK: John Wiley & Sons

Chuong 83 C.J. Chuong, Y.C. Fung (1983), Three-dimensional stress distribution in arteries, *J. Biomech. Engng.*, 105, 268–274

Collins 72 R. Collins, W.C. Hu (1972), Dynamic constitutive relations for fresh aortic tissue, *J. Biomechanics*, 5, 333–337

Comninou 76 M. Comninou and I.V. Yannas (1976), Dependence of stress-strain non-linearity of connective tissues on the geometry of collagen fibers, *J. Biomechanics*, 9, 427–433

Crisp 72 J.D.C. Crisp (1972), Properties of tendon and skin, in *Biomechanics: Its Foundations and Objectives*, ed. by Y.C. Fung. New York: Prentice-Hall

Danielsen 73 D.A. Danielsen (1973), Human skin as an elastic membrane, *J. Biomechanics*, 6, 539–546

Danielsen 75 D.A. Danielsen, S. Natarajan (1975), Tension field theory and the stress in stretched skin, *J. Biomechanics*, 8, 135–142

Decraemer 80a W.F. Decraemer, M.A. Maes, V.J. Vanhuyse (1980a), An elastic stress-strain relation for soft biological tissues based on a structural model, *J. Biomechanics*, 13, 463–468

Decraemer 80b W.F. Decraemer, M.A. Maes, V.J. Vanhuyse, P. Vanpeperstrate (1980), A non-linear viscoelastic constitutive equation for soft biological tissues based on a structural model, *J. Biomechanics*, 13, 559–564

Demiray 72 H. Demiray (1972), A note on the elasticity of soft biological tissues, *J. Biomechanics,* 5, 309–311

Demiray 76 H. Demiray (1976), Some basic problems in biophysics, *Bull. Math. Biol.*, 38, 701–712

Diamant 72 J. Diamant, A. Keller, E. Baer, M. Lii, R.G.C. Arridge (1972), Collagen, ultrastructure and its relations to mechanical properties as a function of ageing, *Proc. R. Soc. London,* B. 180, 293–315

Elden 68 J.R. Elden (1968), Physical properties of collagen fibers, *Int. Rev. Conn. Tiss. Res.*, 4, 283–348

Feit 79 T.S. Feit (1979), Diastolic pressure-volume relations and distribution of pressure and fiber extension accross the wall of a model left ventricle, *Biophys. J.,* 28, 143–166

Frisen 69a M. Frisen, M. Magi, L. Sonnerup, A. Viidik (1969), Rheological analysis of soft collagenous tissues – Part I: Theoretical considerations, *J. Biomechanics*, 2, 13–20

Frisen 69b M. Frisen, M. Magi, L. Sonnerup, A. Viidik (1969), Rheological analysis of soft collagenous tissues – Part II: Experimental evaluations and verifications, *J. Biomechanics*, 2, 21–28

Fung 67 Y.C. Fung (1967), Elasticity of soft tissues in simple elongation, *Am. J. Physiol.,* 213, 1532–1544

Fung 72 Y.C. Fung (1972), Stress-strain history relations of soft tissues in simple elongation, in *Biomechanics: Its Foundations and*

Objectives, Y.C. Fung, N. Perrone, M. Anliker. Englewood Cliffs, NJ: Prentice-Hall

Fung 87 Y.C. Fung (1987), Mechanics of soft tissues, in *Handbook of Bioengineering*, ed. by R. Skalak, S. Chien. New York: McGraw-Hill

Fung 93 Y.C. Fung (1993), *Biomechanics: Mechanical Properties of Living Tissues.* Berlin: Springer-Verlag

Gallagher 82 R.H. Gallagher, B.R. Simon, P.C. Johnson, J.F. Gross (eds.) (1982), *Finite Elements in Biomechanics.* Chichester, UK: John Wiley & Sons

Galford 70 J.E. Galford, J.H. McElhaney (1970), A viscoelastic study of scalp, brain and dura, *J. Biomechanics*, 3, 211–221

Glantz 74 S.A. Glantz (1974), A constitutive equation for the passive properties of muscle, *J. Biomechanics* , 7, 137–145

Glantz 77 S.A. Glantz (1977), A three-element description for muscle with viscoelastic passive elements, *J. Biomechanics* , 10, 5–20

Glaser 65 A.A. Glaser *et al.*. (1965), Refinements of the methods for the measurement of the mechanical properties of unwounded and wounded skin, *Med. Electron. Biol. Engng.,* 3, 411–419

Gou 70 P.F. Gou (1970), Strain energy functions for biological tissues, *J. Biomechanics*, 3, 547–550

Guccione 91 J.M. Guccione, A.D. McCulloch (1991), Passive material properties of intact ventricular myocardium determined from a cylindrical model, *J. Biomech. Engng.*, 113, 42–55

Haut 69 R.C Haut, R.W. Little (1969), The rheological properties of canine anterior cruciate ligaments, *J. Biomechanics*, 2, 289–298

Haut 72 R.C. Haut, R.W. Little (1972), A constitutive equation for collagen fibers, *J. Biomechanics*, 5, 423–430

Hirsch 68 C. Hirsch, L. Sonnerup (1968), Macroscopic Rheology in Collagen Material, *J. Biomechanics*, 1, 13–18

Horowitz 88a A. Horowitz, I. Sheinman, Y. Lanir, M. Perl, S. Sideman (1988), Non-linear incompressible finite element for simulating loading and unloading of cardiac tissue – Part I: Two-dimensional formulation for thin myocardial strips, *J. Biomech. Engng.*, 110, 57–62

Horowitz 88b A. Horowitz, I. Sheinman, Y. Lanir (1988), Non-linear incompressible finite element for simulating loading and unloading of cardiac tissue – Part II: Three-dimensional formulation for thick ventricular wall segments, *J. Biomech. Engng.*, 110, 62–68

Horowitz 88c A. Horowitz, I. Sheinman, Y. Lanir, M. Perl, R.K. Strumpf (1988), Structural three-dimensional constitutive law for the passive myocardium, *J. Biomech. Engng.*, 110, 200–207

Huang 90 X. Huang, M.M. Black, I.C. Howard, E.A. Patterson (1990), A two-dimensional finite element analysis of a bioprosthetic heart valve, *J. Biomechanics*, 23, 753–762

Humphrey 87 J.D. Humphrey, F.C.P. Yin (1987), On constitutive relations and finite deformations of passive cardiac tissue – Part I: A pseudo-strain energy function, *J. Biomech. Engng.*, 109, 298–304

Humphrey 90 J.D. Humphrey, RK Strumpf, F.C.P. Yin (1990), Determination of a constitutive relation for passive myocardium – Parts I and II, *J. Biomech. Engng.*, 112, 333-346

Huygue 91 J. Huygue, M. Dick, H. Van Campen, A. Theo, R.M. Heerthaars (1991), The constitutive behaviour of passive heart muscle tissue: a quasilinear viscoelastic formulation, *J. Biomechanics*, 24, 841–849

Jamison 68 C.E. Jamison, R.D. Marangoni, A.A. Glaser (1968), Viscoelastic properties of soft tissue by discrete model characterization, *J. Biomechanics*, 1, 33–46

Janz 74 R.F. Janz, R.K. Bruce, T.F. Moriarty (1974), Deformation of the diastolic left ventricle – Part II: Non-linear geometric effects, *J. Biomechanics*, 7, 509–516

Jenkins 74 R.B. Jenkins, R.W.M. Little (1974), A constitutive equation for parallel-fibered elastic tissue, *J. Biomechanics*, 7, 397–402

Johnson 92 G.A. Johnson, K.R. Rajagopal, S.L-Y Woo (1992), A single integral finite strain (SIFS) model of ligaments and tendons, *Advances in Bioengineering*, 22, 245–248

Karakaplan 80 A.D. Karakaplan, M.P. Bienek, R. Skalak (1980), A mathematical model of lung parenchyma, *J. Biomech. Engng.*, 102, 124–136

Kastelic 80 J. Kastelic, I. Palley, E. Baer (1980), A structural mechanical model for tendon crimping, *J. Biomechanics*, 13, 887–893

Kenedi 64 R.M. Kenedi, T. Gibson, C.H. Daly (1964), Bioengineering studies of the human skin, in *Biomechanics and Related Bioengineering Topics*, ed. by R.M. Kenedi. Oxford: Pergamon Press

Klosner 69 J.M. Klosner, A. Segal (1969), Mechanical characterization of a natural rubber, *PIBAL Rep.* 69–42, Polytechnic Institute of Brooklyn, New York

Kwan 89 M.K. Kwan, S L-Y. Woo (1989), A structural model to describe the non-linear stress-strain behavior for parallel-fibered collagenous tissues, *J. Biomech. Engng.*, 111, 361–363

Kwan 93 M.K. Kwan, T.H-C. Lin, S.L-Y. Woo (1993), On the viscoelastic properties of the anteromedial bundle of the anterior cruciate ligament, *J. Biomechanics*, 26, 4–5, 447–452

Lanir 79 Y. Lanir (1979), The rheological behavior of the skin: experimental results and a structural model, *Biorheology*, 16, 191–202

Lanir 83a Y. Lanir (1983), Constitutive equations for fibrous connective tissues, *J. Biomechanics*, 16, 1–12

Lanir 83b Y. Lanir (1983), Constitutive equations for the lung tissue, *J. Biomech. Engng.*, 105, 374–380

Lee 75 G.C. Lee, A. Frankus (1975), Elasticity properties of lung parenchyma derived from experimental distorsion data, *Biophys. J.*, 15, 481–493

Lee 82 G.C. Lee, N.T. Tseng (1982), Finite element analysis in soft tissue mechanics, in *Finite Elements in Biomechanics*, ed. by R.H. Gallagher, B.R. Simon, P.C. Johnson, J.F. Gross. Chichester, UK: John Wiley & Sons

Lin 87 H.C. Lin, M.H.W. Kwan, S.L-Y. Woo (1987), On the stress relaxation properties of anterior cruciate ligament (ACL) and the patellar tendon (PT), *Transactions of the Orthopaedic Research Society*, 13, 81

Lyon 88 R.M. Lyon, H.C. Lin, M.K. Kwan, W. Hollis, J.M. Akeson, S.L-Y. Woo (1988), Stress relaxation of the anterior cruciate ligament (ACL) and the patellar tendon (PT), *Transactions of the Orthopaedic Research Society,* 13, 81

Maes 89 M. Maes, V.J. Vanhuyse, W.F. Decraemer, E.R. Raman (1989), A thermodynamically consistent constitutive equation for the elastic force-length relation of soft biological materials, *J. Biomechanics*, 22, 1203–1208

Mooney 40 M. Mooney (1940), A theory of large elastic deformation, *J. Appl. Phys.*, 11, 582–92

Morgan 60 F.R. Morgan (1960), The mechanical properties of collagen fibres: stress-strain curves, *J. Soc. Leath. Trades Chem.*, 44, 170–182

Needleman 83 A. Needleman, S.A. Rabinowitz, D.K. Bogen, T.A. McMahon (1983), A finite element model of the infarcted left ventricle, *J. Biomechanics*, 16, 45–58

Nevo 89 E. Nevo, Y. Lanir (1989), Structural finite deformation model of the left ventricle during diastole and systole, *J. Biomech. Engng.*, 111, 342–349

Novak 94 V.P. Novak, F.C.P. Yin, J.D. Humphrey (1994), Regional mechanical properties of passive myocardium, *J. Biomechanics*, 27, 4, 403–412

Ogden 84 R.W. Ogden (1984), *Non-linear Elastic Deformations*, Chichester: Ellis Horwood / New York: John Wiley & Sons

Peng 78 S.T.J. Peng, R.F. Landel, G.S. Brody (1978), In vitro study of human skin rheology, *Proc. 6Th N.E. Bioeng. Conf.*, 350–354

Pinto 73 J.G. Pinto, Y.C. Fung (1973), Mechanical properties of the heart muscle in the passive state, *J. Biomechanics,* 6, 597–616

Pioletti 97 D.P. Pioletti (1997), Viscoelastic properties of soft tissues: application to knee ligaments and tendons, Ph.D. Thesis, EPFL – Lausanne

Pradas 90 M.M. Pradas, R.D. Calleja (1990), Non-linear viscoelastic behavior of the flexor tendon of the human hand, *J. Biomechanics*, 23, 773–781

Purslow 89 P.P. Purslow (1989), Strain induced reorientation of an intramuscular connective tissue network: Implications for passive muscle elasticity, *J. Biomechanics*, 22, 21–31

Ridge 66 Ridge (1966), Mechanical properties of skin: A bioengineering study of skin texture, *J. Appl. Physiol.,* 21, 1602–1606

Sacks 93 M.S. Sacks, C.J. Chuong (1991), A constitutive relation for passive right-ventricular free wall myocardium, *J. Biomechanics*, 26, 1341–1345

Sakata 72 K. Sakata, G. Parfitt, K.L. Pinder (1972), Compressive behavior of physiological tissue, *Biorheology*, 9, 173–184

Sanders 73 R. Sanders (1973), Torsional elasticity of human skin in vivo, *Pflügers Arch.,* 342, 255–260.

Sanjeevi 82 R. Sanjeevi, N. Somanathan, D. Ramaswamy (1982), A viscoelastic model for collagen fibres, *J. Biomechanics,* 15, 181–183

Sedlin 66 E.D. Sedlin, L. Sonnerup (1966), Rheological considerations in the physical properties of bone, *3rd European Symp. Calcified Tissues*, ed. by H. Fleisch, H. J. J Blackwood, M. Owen. Berlin: Springer-Verlag

Shoemaker 86 P.A. Shoemaker, D. Scheider, M.C. Lee, Y.C. Fung (1986), A constitutive model for two-dimensional soft tissues and its application to experimental data, *J. Biomechanics*, 19, 6, 695–702

Silver 87 F.H. Silver (1987), *Biological materials: structure, mechanical properties and modeling of soft tissues.* New York: New York University Press

Skalak 82 R. Skalak, S. Chien (1982), Rheology of blood cells as soft tissues,*4th Int. Cong. Biorh. Symp. Mech. Prop. Liv. Tiss.*, 453–461

Snyder 72 R.W. Snyder (1972), Large deformation of isotropic biological tissue, *J. Biomechanics*, 5, 601–606

Snyder 75 R.W. Snyder, L.H.N. Lee (1975), Experimental study of biological tissue subjected to pure shear, *J. Biomechanics*, 8, 415–419

Tong 76 P. Tong, Y.C. Fung (1976), The stress-strain relationship for the skin, *J. Biomechanics*, 9, 649–657

Tozeren 83 A. Tozeren (1983), Static analysis of the left ventricle, *J. Biomech. Engng.*, 105, 39–46

Treloar 75 L.R.G. Treloar (1975), *The physics of rubber elasticity*, Oxford: Clarendon Press

Trevisan 83 Trevisan (1983), Etude des propriétés rhéologiques des tissus biologiques – Application au comportement mécanique des ligaments naturels et artificiels, Ph.D. Thesis, Paris XII

Vawter 80 D.L. Vawter (1980), A finite element model for macroscopic deformation of the lung, in *Finite Elements in Biomechanics*, ed. by R.H. Gallagher, B.R. Simon, P.C. Johnson, J.F. Gross. Chichester, UK: John Wiley & Sons

Veronda 70 D.R. Veronda, R.A. Westmann (1970), Mechanical characterization of skin-finite deformations, *J. Biomechanics*, 3, 111–124

Viidik 68 A. Viidik (1968), A rheological model for uncalcified parallel-fibered collagenous tissue, *J. Biomechanics*, 1, 3–11

Viidik 80 A. Viidik, J. Vuust (1980), *Biology of collagen : proceedings of a symposium*, Aarhus, July –August 1978. London: Academic Press

Viidik 87 A. Viidik (1987), Properties of tendons and ligaments, in *Handbook of Bioengineering*, ed. by R. Skalak, S. Chien, New York: McGraw-Hill

Vito 73 R.P. Vito (1973), A note on Arterial Elasticity, *J. Biomechanics*, 6, 561–564

Vito 80 R.P. Vito, J. Hickey (1980), The mechanical properties of soft tissues – Part II: The elastic response of arterial segments, *J. Biomechanics*, 13, 951–957

Vlasblom 67 D.C. Vlasblom (1967), *Skin Elasticity*, Ph.D. Thesis, University of Utrecht, The Netherlands

Wertheim 47 M.G. Wertheim (1847), Mémoire sur l'élasticité et la cohésion des principaux tissus du corps humain, *Annls. Chim. Phys.*, 21, 385–414

Wijn 76 P.F. Wijn, A.J.M. Brakkee, G.J.M. Stienen, A.J.H Vendrick (1976), Mechanical properties of human skin in-vivo for small deformations: a comparison of uniaxial strain and torsion measurements, in *Bedsore Biomechanics*, ed. by R.M. Kenedi, J.M. Cowden, J.T. Scales. London: Macmillan

Wong 71 Wong, A.Y.K. (1971), Mechanics of cardiac muscle, based on Huxley's model: mathematical simulation of isometric contraction, *J. Biomechanics,* 4, 529–540

Woo 76 S.L-Y. Woo, W.H. Akeson, G.F. Jemmott (1976), Measurements of non-homogeneous, directional mechanical properties of articular cartilage in tension, *J. Biomechanics*, 9, 785–791

Woo 82 SL-Y. Woo (1982), Mechanical properties of tendons and ligaments – Parts I and II, *4th Int. Cong. Biorh. Symp. Mech. Prop. Liv. Tiss.*, 385–408

Woo 93 S-Y Woo, G.A. Johnson, BA Smith (1993), Mathematical modeling of ligaments and tendons, *J. Biomech. Engng.*, 115, 468–473

Yang 91 M. Yang, L.A. Taber (1991), The possible role of poroelasticity in the apparent viscoelastic behavior of passive cardiac muscle, *J. Biomechanics*, 24, 7, 587–597

Zeng 87 Y.J. Zeng, D. Yager, Y.C. Fung (1987), Measurement of the mechanical properties of the human lung tissue, *J. Biomech. Engng.,* 109, 160–174

5 Muscle Contraction Modeling

Soft tissue constitutive modeling requires a particular investigation into muscle contraction since muscles also exhibit an active behavior. In this area, there are three major approaches corresponding to different purposes. Most models are devoted to muscle force prediction, and only provide models for the global uniaxial output force of given muscles in defined conditions and experiments. These models don't take into account the local mechanics involved inside the fibers. Conversely, some models are devoted instead to the understanding of the contractile mechanism, and describe the chemico-mechanical aspects of the contraction process at the sarcomeral level, but have hardly been related to a realistic global output force involving the 3D anatomical and passive properties of muscle. Only few studies attempt to provide a model of muscle including anatomical and mechanical, active and passive properties, allowing a realistic simulation of its contractile behavior in relation with its deformation and its global output force. These different aspects are overviewed in the following.

5.1 Different Models for Different Purposes

5.1.1 Contraction Kinematics Modeling

Deep investigations have been led towards the identification and description of the complex sliding filament mechanism involved in muscle contraction at the sarcomeral level. A fundamental incursion in the contractile mechanism has been led by Huxley (Huxley 57), who first formulated the cross-bridge and sliding filament theory. Figure 5.1 gives a kinematic description of the mechanism as illustrated by Huxley et al. (Huxley 71). According to Huxley's theory, force is generated by the sarcomeres in response to activation by means of molecular

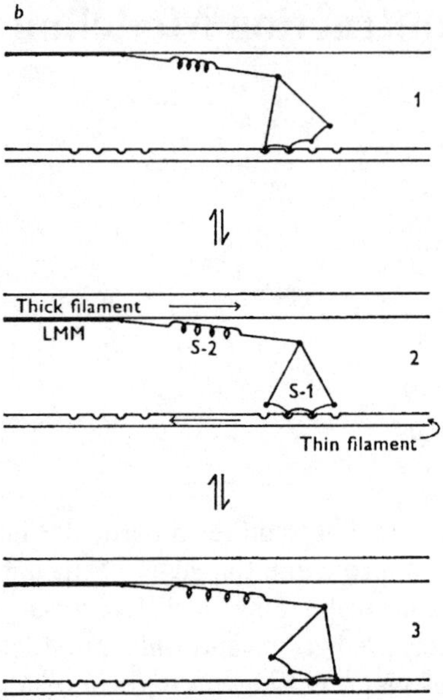

Fig. 5.1. Kinematic model for force generation
(reprinted from (Huxley 74) with permission of Cambridge University Press)

linkages, called *cross-bridges*, between thin *actin* and *myosin* filaments (Sect. 1.2.1). Noting N the number of *myosin* heads in a half sarcomere, and n the density of cross-bridges, the number of free *myosin* heads is $(N-n)$. The kinematics of the cross-bridge formation may be expressed as:

$$\begin{cases} \dfrac{dn}{dt} = (N-n)\, f(s,x,t) - n\, g(s,x,t) \\ \dfrac{dn}{dt} = \dfrac{\partial n}{\partial t} + v_s \dfrac{\partial n}{\partial x} \end{cases} \quad \text{with} \quad v_s = \frac{dx}{dt} \tag{5.1}$$

where s is the sarcomere length, x the cross-bridge length, t the time
v_s the shortening velocity, f and g the cross-bridge formation rates

After solving (5.1) for n, the sarcomere force may be formulated as:

$$F_s(t) = \int_{-\infty}^{+\infty} K\, n(x,t)\, x\, dx \qquad \text{K: cross-bridge elastic stiffness} \tag{5.2}$$

Further analysis followed, on the basis of Huxley's description, notably with Wong et al., who lead a mathematical simulation of heart contractions (Wong 71), with Van der Broek et al., who attempted to model the beating of the left ventricle (Broek 80), with Wood et al., who proposed a model for the mechanical transients of contraction (Wood 81), with Zahalak et al. who formulated constitutive relations based on the cross-bridge kinetics (Zahalak 90), and with Ma et al., who suggested a distribution moment model for activation and contraction (Ma 91). These microscopic aspects, however, are too complex for the macroscopic properties aimed at in our plan. We therefore investigate this area no further.

5.1.2 Symbolic Muscle Modeling

Motion analysis usually considers the influence of a group of muscles on the motion of a joint, with no reference to the muscular anatomy, or with reference to an equivalent muscle representing several associated or antagonist muscles. This approach may be illustrated by Baildon's et al. muscle model in which the muscular actions are described in terms of the resulting torque observed around the elbow (Fig. 5.2) (Baildon 83). Some other models of muscle force may be derived from the analysis of the global mechanical properties around a joint under specific experimental conditions. For example, Niku et al. presented a model for evaluating the viscosity of the flexor muscles around the elbow joint (Fig. 5.3) (Niku 89). As such approaches do not really model the contraction of individual muscles in taking into account their real anatomy, we do not analyze them.

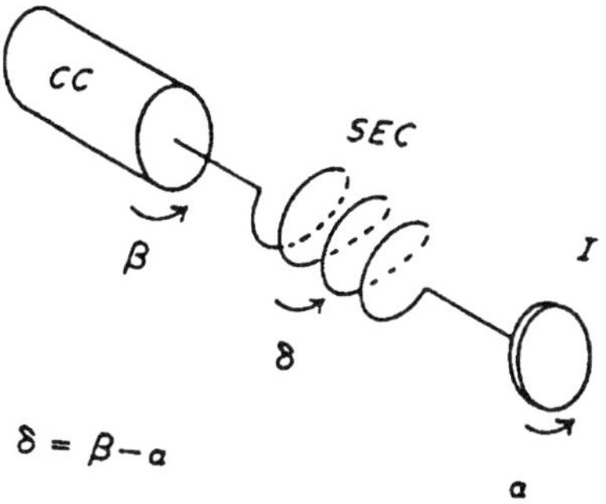

Fig. 5.2. Baildon's muscle model for elbow flexion
(reprinted from (Baildon 83)

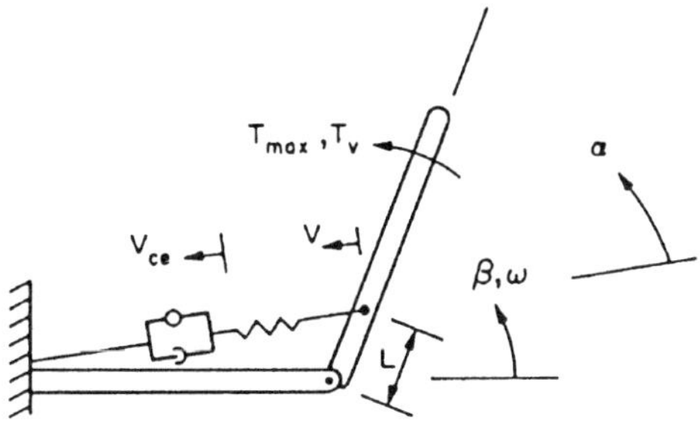

Fig. 5.3. Niku's muscle model for elbow flexion (reprinted from (Niku 89) with permission of Elsevier Science)

5.1.3 Force Prediction from PCSA

The maximal tensile strength of a muscle is usually estimated by means of *its physiological cross-sectional area*, called PCSA (Winter 90). The physiological cross-sectional area allows the estimation of the force/stress developed by the muscle taking into account the angle of pennation of the fibers. This practice is based on the assumption that the bigger a muscle is, the stronger it is. The PCSA corresponds to the area of the muscle cross-section actually sustaining the actuator tension. It is defined as the projection of the muscle cross-section on a plane perpendicular to the muscle force direction. Therefore, for a straight musculotendon actuator, the PCSA corresponds to the real section of the muscle, whereas for a bipennate muscle, it is equal to the real section reduced by the cosine of the pennation angle (Fig. 5.4). For example, the PCSA of an imaginary muscle whose contractile fibers are orthogonal to the tendon fibers, (i.e., a pennation angle of 90°) is null, which meaning that the muscle contraction does not contribute to the longitudinal force of the muscle (Sect. 1.2.1). However, except for muscles with strong pennation, the pennation angle is rarely taken in account and most PCSA are estimated with the following relation:

$$\mathrm{PCSA} = \mathrm{S}_0^{\mathrm{M}} = \frac{\mathrm{V}_0^{\mathrm{M}}}{\mathrm{l}_0^{\mathrm{M}}} \tag{5.3}$$

where $\mathrm{S}_0^{\mathrm{M}}$ is the estimated cross-sectional resting area of the muscle
$\mathrm{l}_0^{\mathrm{M}}$ is the measured resting length of the muscle
$\mathrm{V}_0^{\mathrm{M}}$ is the measured resting volume of the muscle

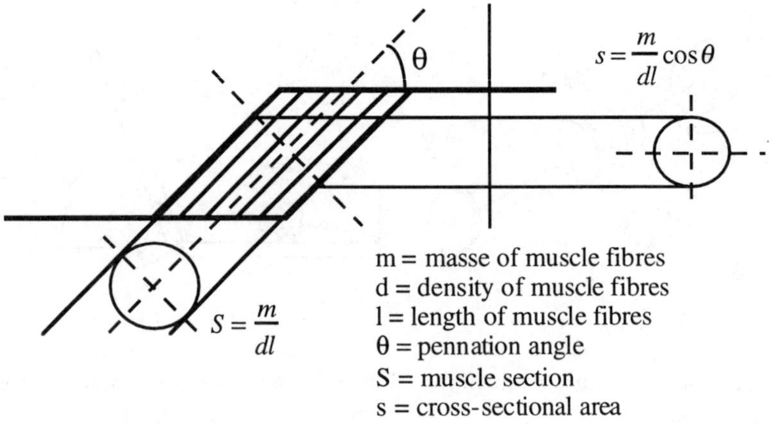

Fig. 5.4. Muscle physiological cross-sectional area

Assuming therefore a constant material strength factor T^M, the PCSA may be used to estimate, by means of the following relation, the maximum force F^M_{max} a muscle may develop:

$$F^M_{max} = PCSA \times T^M \qquad \text{with} \qquad T^M = 0.4 - 1.0\ \text{MPa} \tag{5.4}$$

where T^M corresponds to the Lagrange stress. Muscle force prediction represents an important area of investigation for purpose of joint simulations and motion analysis. For example, Karlsson et al. used $T^M = 0.7$ MPa for force prediction of muscles around the shoulder (Karlsson 92).

5.1.4 Force Estimation from EMG

Two techniques are available in practice to estimate the activation rate of a muscle actuator: the electromyographic measurements (EMG) and the intramuscular pressure measurements (IMP). Both methods provide complementary information on the muscle activation state. However, as they provide more accurate data than IMP measurements, EMG measurements have been more commonly used. The relationship between rectified smoothed EMG and isometric muscle force has been reported to be linear (Sect. 1.2.3) (Hof 77) of the form:

$$F^M = \alpha A + \beta \tag{5.5}$$

where
F^M is the muscle contractile force
A is the amplitude of the mean rectified EMG signal
α is the gain factor and β is a constant

The model has later been observed to be extendable to dynamic contractions (Hof 81a-d). However, there are trustworthy investigators who consistently found nonlinear relations (Hof 88). Other factors must be borne in mind when considering the EMG-force relationship such as the variation between individuals and the size of the muscle under investigation (Cutts 93). Nevertheless, the assumption of a linear EMG to force relationship is very common, and may be justified by the fact that in the functional range, muscles probably do not reach extreme lengths and carry only moderate loads. However, EMG data must be undertaken only with extreme caution due to technical limitations and misinterpretations of the results (Cutts 93).

Bobet et al. also investigated the muscle force response with respect to time and excitation. They proposed to model the isometric muscle force response to impulse as the response of a critically damped, linear second-order system. Since linearity is assumed, the response to a train of pulses is the sum of the responses to each stimulus in the train. The response takes the form of a differential relation:

$$\frac{d^2F(t)}{dt^2} + A(t)\frac{dF(t)}{dt} + B(t)\,F(t) = \sum_{i=0}^{n} C_i(t)\,\delta(t - t_i) \qquad (5.6)$$

This relation may be solved for each pulse under defined limits, and leads to a force prediction model in exponential form (Bobet 93).

5.1.5 Constitutive Modeling

The ultimate step in muscle modeling is the integration of the active behavior with the passive behavior into a same constitutive relationship. Few studies have addressed this problem. Taking into account the series combination of the contractile element with the series element in Hill's model (Fig. 1.10), Pinto et al. proposed to express the strength of the series element in terms of the contraction of the contractile element, introducing therefore the active behavior in the expression of the global muscle force as:

$$T^M(\lambda, t) = T^{PE}(\lambda) + T^{SE}(\lambda, t) \qquad T^{SE} = T^{CE} = A(\lambda)\, t^v e^{-\delta t} \qquad (5.7)$$

with:
T^M is the Lagrangian stress of the muscle
T^{PE} is the Lagrangian stress of the parallel element
T^{SE} is the Lagrangian stress of the serial element
T^{CE} is the Lagrangian stress of the contractile element
λ is the muscle extension ratio, t is the time
A is the amplitude of the contraction force function of λ
v is the contraction delay factor, δ is the muscle anisotropy factor

This model has been developed to describe the active behavior of papillary muscles (Pinto 87). This expression could fit the experimental data on a length tension relationship, time course in single twitches, a quick-stretch response, and a quick-release response. Extension to isovolumic contraction was accomplished by a different set of constants of A, v, δ.

However, extrapolation of one-dimensional data to three dimensions is not possible for large deformations. For this reason, Taber et al. attempted to combine the active and passive properties of muscle within a same strain energy function:

$$W = W^m\left(I_1^m\right) + W^f\left(I_1^f\right) + W^a\left(I_1^a\right) \qquad \text{where} \qquad (5.8)$$

$$W^m\left(I_1^m\right)=C\left[e^{b_0\left(I_1^m-3\right)}-1\right]$$ is the isotropic matrix contribution

$$W^f\left(I_1^f\right)=Ca_1\left[e^{b_1\left(I_1^f-1\right)}-b_1I_1^f-1+b_1\right]$$ is the muscle fiber passive contribution

$$W^a\left(I_1^a\right)=C\,a_2(t)\left(I_1^a+I_1^{a^{-1}}-2\right)^{b_2(t)}$$ is the muscle fiber active contribution

where I_1^m is the first strain invariant of the isotropic matrix
I_1^f is the first strain invariant of the passive muscle fibers
I_1^a is the first strain invariant of the active muscle fibers
C, a_1, b_1 and b_0 are passive material constants
a_2 and b_2 are the active time-varying coefficients
t is the time from the onset of activation

The model has been applied to canine myocardium characterization (Taber 91a-b). However, the authors reported that due to lack of biaxial data for active muscle, the active functions a_2 and b_2 could not be uniquely determined.

5.2 Musculotendon Dynamics Modeling

5.2.1 Force-Length Models

Biomechanical motion analysis improve in realism when the anatomical topology of the muscles is taken into account. The muscle actions on the skeleton are usually modeled by one or several lines of action. The computation of their forces requires the availability of realistic muscle force models. For this purpose, Otten suggested to fit the active length-tension curve with the relation (Otten 87):

$$F=e^{-\left[\frac{(\varepsilon+1)^{\beta}-1}{w}\right]^{\rho}} \tag{5.9}$$

where F is the normalized active force of a muscle actuator
ε is the exponential function of the muscle strain
β is the skewness, ρ the roundness, w the width factors

The model was extended to muscles composed of several independent spindles:

$$F = \sum_j K_j \, e^{-\left[\frac{(\varepsilon+1)^{\beta_j}-1}{w}\right]^{\rho_j}} \tag{5.10}$$

where K_j represents the contribution of each action line $\mathbf{j}$ to the total force F. This generalization has been used by Baratta et al. for the characterization of nine different skeletal muscles from cat (Baratta 93).

Otten's relation was also improved by Kaufman et al. to account for muscle pennation by mean of an index of muscle fiber architecture i_a. The index of architecture was defined as the ratio of the mean fiber length to the muscle optimum length (Fig. 5.5), and incorporated in Otten's relation as :

$$F = e^{-\left[\frac{(\varepsilon+1)^{\beta}-1}{w}\right]^{\rho}} \quad \text{with} \quad \begin{cases} \rho = 2 \\ \beta = 0.96343\left(1 - \dfrac{1}{i_a}\right) \\ w = 0.35327\,(1 - i_a) \end{cases} \quad \text{for } i_a < 1 \tag{5.11}$$

$$F = e^{-[2.727 \times \ln(\varepsilon+1)]^2} \qquad \text{for } i_a = 1 \tag{5.12}$$

They have applied this model to force prediction around the elbow (Kaufman 89).

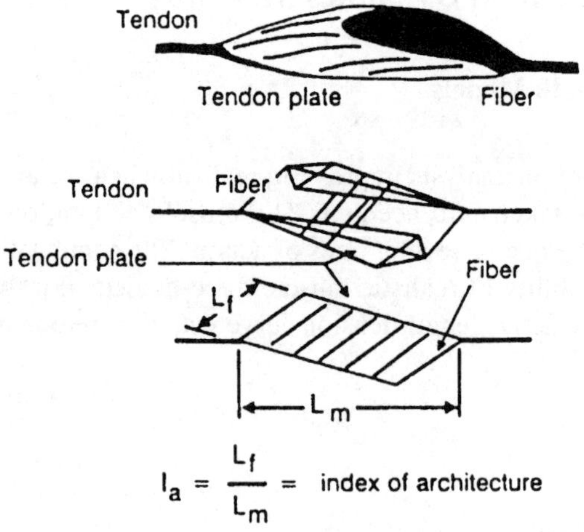

Fig. 5.5. Index of architecture for a pennated muscle
(reprinted from (Kaufman 89) with permission of Elsevier Science)

5.2.2 Force-Length-Velocity-Activation Relation Description

Zajac et al. led a complete analysis of the different parameters affecting the musculotendon force development, in order to provide a method to fully describe and compute the contraction forces of real actuators. In a musculotendon ensemble, the force is developed in the muscle and transmitted to the bones through the tendons. The true contraction force is thus somewhat filtered by the non-linear passive properties of the connective tissues. It also depends non-linearly on the current length, velocity and activation of the muscle spindle. Furthermore, the muscle activation is controlled by the neural excitation by means of chemico-mechanical interactions which determine the activation dynamics. Zajac modeled the actuator as composed a linear spring for tendon, in pennation with a Hill type model of muscle (Fig. 5.6): a series association of a contractile element and a non-linear spring, in parallel with another non-linear spring (Sect. 1.2.2) (Zajac 86). From the geometry of the model, the following statements arise:

$$\cos\alpha = \sqrt{1-\left(\frac{W}{l^M}\right)^2} \qquad \text{where W is assumed constant} \tag{5.13}$$

$$l^M = \frac{W}{\sin\alpha} \qquad v^M = \frac{dl^M}{dt} = -\frac{W}{\sin^2\alpha}\cos\alpha\frac{d\alpha}{dt} \tag{5.14}$$

$$l^{M\alpha} = l^M\cos\alpha = \frac{W}{\tan\alpha} \qquad v^{Ma} = \frac{dl^{M\alpha}}{dt} = -\frac{W}{\sin^2\alpha}\frac{d\alpha}{dt} \tag{5.15}$$

$$l^{MT} = l^T + l^M\cos\alpha \qquad v^{MT} = \frac{dl^{MT}}{dt} = v^T + \frac{v^M}{\cos\alpha} \tag{5.16}$$

From the structure of the model also come the following relations:

$$P^T = P^M\cos\alpha \qquad \frac{dP^T}{dt} = \frac{dP^M}{dt}\cos\alpha - P^M\sin\alpha\frac{d\alpha}{dt} \tag{5.17}$$

$$l^M = l^{PE} = l^{CE} + l^{SE} \qquad P^M = P^{PE} + P^{SE} = k^{PE}\Delta l^M + k^{SE}\Delta l^{SE} \tag{5.18}$$

$$v^M = v^{PE} = v^{CE} + v^{SE} \qquad \frac{dP^M}{dt} = \left(k^{PE} + k^{SE}\right) v^M - k^{SE}v^{CE} \tag{5.19}$$

Then setting $k^M = k^{PE} + k^{SE}$ and $k^{Ma} = k^M\cos\alpha + \left(\frac{P^T}{l^M}\right)\tan^2\alpha$ (5.20)

(5.17) may be expressed in the form:

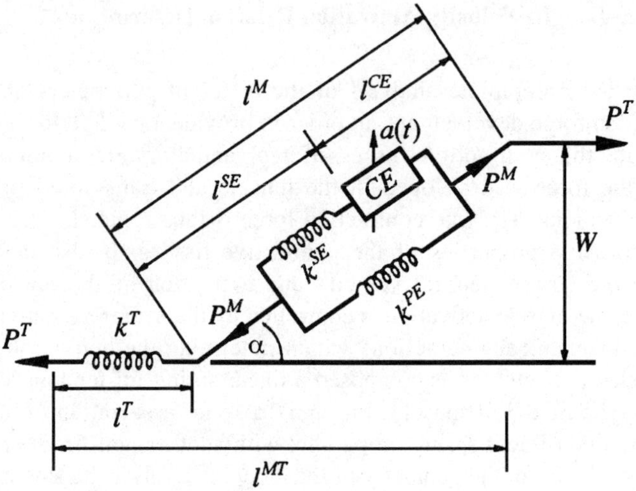

Fig. 5.6 Zajac's musculotendon model
(reprinted from (Pandy 90) with permission of Elsevier Science)

$$\frac{dP^{T}}{dt} = k^{M\alpha}\cos\alpha\left[\frac{v^{M}}{\cos\alpha} - \frac{k^{SE}}{k^{M\alpha}}v^{CE}\right] \tag{5.21}$$

Considering the uniaxial extension, Zajac assumed tendon as a linear elastic spring with stiffness k^{T} and slack length l_0^{T} such that:

$$P^{T} = k^{T}\left(l^{T} - l_0^{T}\right) \qquad \frac{dP^{T}}{dt} = k^{T}v^{T} \tag{5.22}$$

$$v^{MT} = \frac{dP^{T}}{dt}\left[\frac{1}{k^{T}} + \frac{1}{k^{M\alpha}\cos\alpha}\right] + \frac{k^{SE}}{k^{M\alpha}}v^{CE} \tag{5.23}$$

leading to the final equation describing the musculotendon actuator dynamics:

$$\frac{dP^{T}}{dt} = \frac{\left(k^{M\alpha}\cos\alpha\right)k^{T}}{\left(k^{M\alpha}\cos\alpha\right) + k^{T}}\left[v^{MT} - \left(\frac{k^{SE}}{k^{M\alpha}}\right)v^{CE}\right] \tag{5.24}$$

The passive stiffness of the parallel element k^{PE} is defined by the experimental force-length curves (Figs. 5.7,5.8), while the stiffness of the serial spring element k^{SE} is assumed to be given by the equation (Pandy 90):

$$k^{SE} = \frac{dP^{SE}}{dl^{SE}} = \frac{\left(100\,P^{SE} + 10\,P_0\right)}{l_0^M} \tag{5.25}$$

in which $P_0 = P_{fa}^{CE}\left(l_0^M\right)$ is the isometric contraction force of the fully activated muscle at its optimal length l_0^M. The muscle stiffness k^M may thus be determined. The tendon force-length relationship and stiffness may be finally be written as:

$$P^T = \left(\frac{P_0}{\varepsilon_0^T}\right)\varepsilon^T \qquad \text{with} \qquad \varepsilon^T = \frac{\left(l^T - l_0^T\right)}{l_0^T} \qquad \text{i.e.:} \tag{5.26}$$

$$P^T = k^T\left(l^T - l_0^T\right) \qquad \text{with} \qquad k^T = \frac{1}{l_0^T}\left(\frac{P_0}{\varepsilon_0^T}\right) \tag{5.27}$$

The last unknown in the model is v^{CE}, the velocity of shortening of the contractile element. Assuming first, uncoupled contraction and activation mechanisms, Zajac expressed the effective isometric contraction force P_{iso}^{CE} as the experimental isometric contraction force of the fully activated muscle $P_{fa}^{CE}\left(l^M\right)$, scaled by its effective activation rate $a(t)$, giving:

$$P_{iso}^{CE} = a(t)\,P_{fa}^{CE}\left(l^M\right) \tag{5.28}$$

Zajac further investigated the muscle activation dynamics and provided a first order differential equation relating the activation rate $a(t)$ of the muscular contraction to the neural excitation signal $u(t)$ of the muscle as (Zajac 89):

$$\frac{da(t)}{dt} + \left[\frac{1}{\tau_{act}}\left[\beta + [1-\beta]\,u(t)\right]\right]a(t) = \frac{1}{\tau_{act}}u(t) \qquad \beta,\ \tau_{act}\ \text{constants} \tag{5.29}$$

In practice, considering the excitation as an on-off signal leads to reduce (5.29) to:

$$\frac{da(t)}{dt} = \frac{1}{\tau_{rise}}(1-a)\ \text{for}\ u(t)=1 \qquad \frac{da(t)}{dt} = \frac{1}{\tau_{fall}}\left(a_{min} - a\right)\ \text{for}\ u(t)=0 \tag{5.30}$$

Further assuming uncoupled isometric force-length and force-velocity relationships, Zajac suggested to normalize the isometric contraction force P_{iso}^{CE} for scaling the force-velocity curve, in order to express the effective contraction force as:

$$P^{CE} = \left[\frac{P^{CE}_{iso}\left(l^M, a(t)\right)}{P_0} \right] P^{CE}\left(v^{CE}\right) \tag{5.31}$$

As a result of this development, the muscle contraction force P^{CE} appears as a function of the length l^M of the muscle, the velocity of shortening v^{CE} and the activation $a(t)$ of the contractile element. It is therefore possible to reverse the relation and extend the velocity of shortening v^{CE} of the contractile element as a function of the musculo-tendon tension P^T, length l^{MT} and activation $a(t)$, as:

$$v^{CE} = v^{CE}\left[P^T, l^{MT}, a(t)\right] \tag{5.32}$$

Referring back to (5.26), the dynamics of the musculo-tendon actuator reduces finally to a function of its tension P^T, length l^{MT}, velocity v^{MT}, activation $a(t)$ as:

$$\frac{dP^T}{dt} = f\left[P^T, l^{MT}, v^{MT}, a(t)\right] \tag{5.33}$$

The corresponding set of parameters needed to describe the actuator are:

P_0: the peak isometric active contraction force. $P_0 = P^{CE}_{fa}\left(l^M_0\right)$

l^M_0: the muscle length at which P_0 may be developed

l^T_0: the tendon slack length

ε^T_0: the tendon strain for $P^T = P_0$

α_0: the pennation angle for $l^M = l^M_0$

together with the experimental isometric force-length and force-velocity curves:

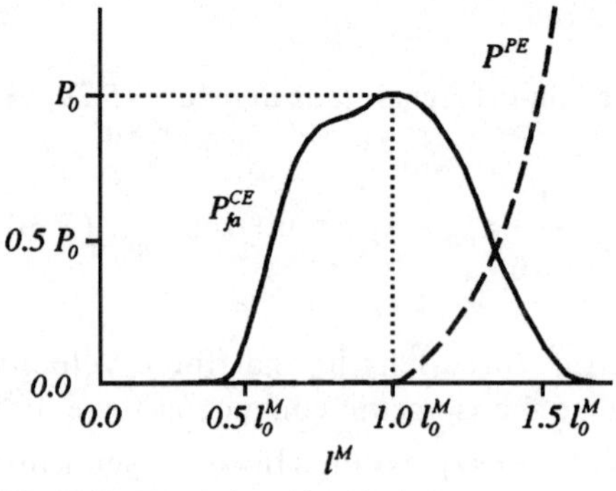

Fig. 5.7 Force-length relations

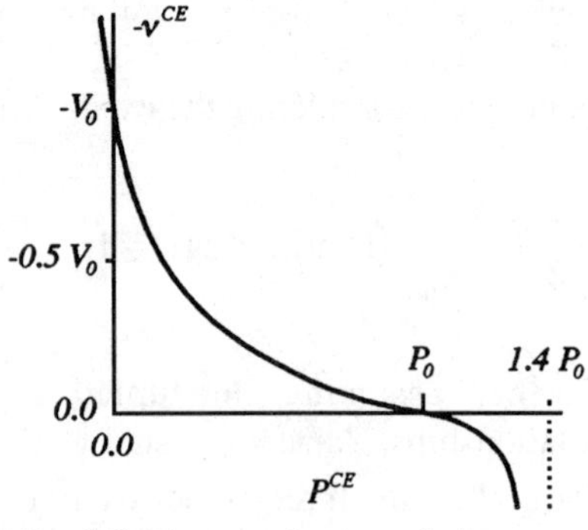

Fig. 5.8 Force-velocity relations

(reprinted from (Pandy 90) with permission of Elsevier Science)

5.3 The Fiber Contraction Force Model

5.3.1 Muscle Geometric Description

Muscle effective modeling requires to take into account its specific contractile properties. The activable stresses induced in the muscle body depending on its length and shortening velocity must be modeled in accounting for their actual fiber-oriented distribution through the volume as well as for their resulting force at the tendon extremity. Two types of muscle force models may be outlined from Sect. 5.1 and 5.2. One considers the macroscopic effects of the contraction at the extremity of the tendon – for example Zajac's model (5.35) (Zajac 86) – while the other one considers the internal contractile mechanism at the sarcomeral level – for example Huxley's model (5.1) (Huxley 57) – and formulates its force on the basis of its kinetics. These models are complementary. In our application, both properties are required. On one hand, the coincidence of the model output force with the measured muscle force is a priority, as basis for skeletal motion simulation, but on the other hand the accurate modeling of the internal contraction stress is the basis for a realistic simulation of the muscle deformation. As far as we have investigated the biomechanics literature, we have not found a muscle model integrating both aspects of the real contraction phenomenon. We therefore suggest an alternative approach closely related to the finite element discretization purpose for modeling the muscle fiber contraction stress including the macroscopic as well as structural properties of the muscle tissue.

Considering the fact that, in muscle, the fibers are organized somehow like a flow, in a way similar to the flow-lines considered in fluid mechanics or in electromagnetism, we may describe the structural properties of muscle by means of a mathematical approach. Muscle may be viewed as a continuous deforming three-dimensional object $\mathcal{M}$, in which each point may be described with respect to the muscle local coordinate system $(\mathbf{x}_1, \mathbf{x}_2, \mathbf{x}_3) \equiv (\mathbf{x}, \mathbf{y}, \mathbf{z})$. The deformation may be described with respect to a global reference frame $(\mathbf{X}_1, \mathbf{X}_2, \mathbf{X}_3)$ corresponding to the reference configuration of the muscle. For a given muscle, the fibrous structure may be described by means of a vector field function $\mathbf{d}$, defining for each point P of the muscle, the unitary tangent vector to the fiber, i.e.:

$$\forall \mathrm{P} \in \mathcal{M} \qquad \mathbf{d}:\ \mathrm{P} \mapsto \mathbf{d}(\mathrm{P}) \qquad (5.34)$$

Considering the different pennate configurations represented in Fig. 1.8, we assume there are globally three types of muscle fiber distribution: *fusiform*, *triangular,* and *spiral.* We suggest simple mathematical descriptions for these configurations.

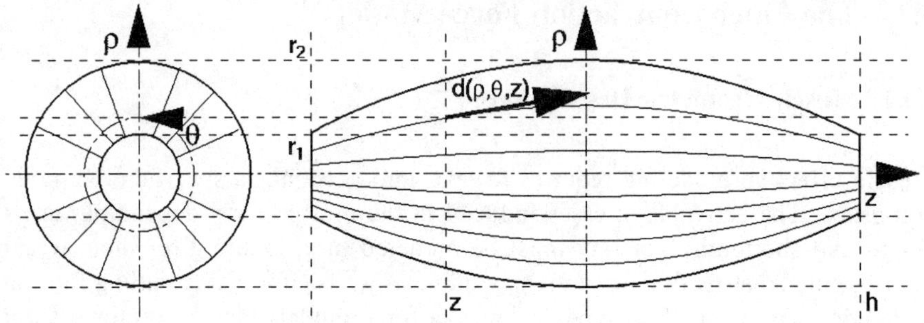

Fig. 5.9. Fusiform muscle description

Fusiform. As shown in Fig. 5.9, the fibers of a fusiform muscle may be represented by parabolic curves, with equation in cylindrical coordinates (ρ, θ, z):

$$\rho(z) = \rho_0\left(1 + a_F z^2\right) \quad \text{for} \quad (\theta, z) \in [0, 2\pi] \times [-h, h] \quad \text{with} \tag{5.35}$$

$$(\rho_0, \theta_0) \equiv \text{fiber constants} \quad (\rho_0, \theta_0) \in [0, r_2] \times [0, 2\pi] \quad a_F = -\frac{r_2 - r_1}{r_2 h^2}$$

Then, the fiber direction at a point $P(\rho, \theta, z)$ may be defined by the function:

$$\mathbf{d}: P(\rho, \theta, z) \mapsto \mathbf{d}(P) = \begin{bmatrix} d_\rho = \dfrac{2 a_F \rho z}{1 + a_F z^2} \\ d_\theta = 0 \\ d_z = 1 \end{bmatrix} \quad \text{with} \quad \rho_0 = \frac{\rho}{1 + a_F z^2} \tag{5.36}$$

Triangular. As shown in Fig. 5.10, the fibers of a triangular muscle may be represented by straight lines with equation in Cartesian coordinates (x, y, z):

$$y(z) = y_0\left(1 + a_T z\right) \quad \text{for} \quad (x, z) \in \left[-\frac{e}{2}, \frac{e}{2}\right] \times [0, h] \quad \text{with} \tag{5.37}$$

$$(x_0, y_0) \equiv \text{fiber constants} \quad (x_0, y_0) \in \left[-\frac{e}{2}, \frac{e}{2}\right] \times [0, r_1] \quad a_T = \frac{r_2 - r_1}{r_1 h}$$

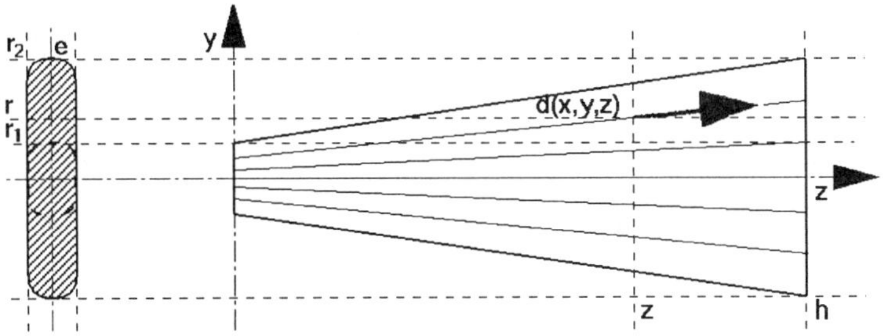

Fig. 5.10. Triangular muscle description

Then, the fiber direction at a point $P(x, y, z)$ may be defined by the function:

$$\mathbf{d}: P(x,y,z) \mapsto \mathbf{d}(P) = \begin{bmatrix} d_x = 0 \\ d_y = \dfrac{a_T y}{1 + a_T z} \\ d_z = 1 \end{bmatrix} \qquad \text{with} \qquad y_0 = \frac{y}{1 + a_T z} \tag{5.38}$$

Spiral. As shown in Fig. 5.11, the fibers of a spiral muscle may be represented by straight lines, with equation in cylindrical coordinates (ρ, θ, z):

$$\begin{cases} \rho = \dfrac{\rho_0}{\cos\theta} \\ z = a_S(\theta - \theta_0) \end{cases} \quad \text{with} \quad \begin{cases} \theta_0 = -\dfrac{\alpha}{2} & \text{for} \quad \theta \in \left[-\dfrac{\alpha}{2}, \dfrac{\alpha}{2}\right] \\ \theta_0 = \pi - \dfrac{\alpha}{2} & \text{for} \quad \theta \in \left[\pi - \dfrac{\alpha}{2}, \pi + \dfrac{\alpha}{2}\right] \end{cases} \tag{5.39}$$

with $\quad a_S = \dfrac{h}{\alpha} \quad$ and $\quad \rho_0 \in [0, r] \quad \rho_0 \equiv$ fiber constant

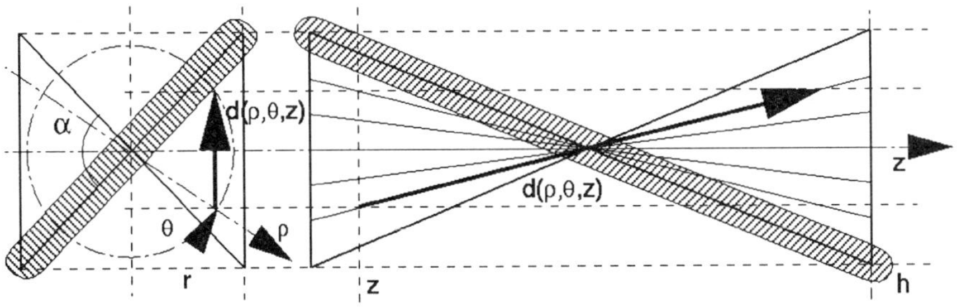

Fig. 5.11. Spiral muscle description

Then, the fiber direction at a point $P(\rho,\theta,z)$ may be defined by the function:

$$\mathbf{d}:\ P(\rho,\theta,z) \mapsto \mathbf{d}(P) = \begin{bmatrix} d_\rho = \rho\tan\theta \\ d_\theta = \rho \\ d_z = a_S \end{bmatrix} \qquad \text{with} \qquad \rho_0 = \rho\cos\theta \tag{5.40}$$

5.3.2 The Contraction Force Function

In a muscle, each fiber may have different strength (resp. cross-section) and different activation rate at a given time. As explained in Sect. 1.2, the smaller units are recruited before the larger ones, and released in the reverse order. The inner fibers are usually powerful, while the outer ones rather stand for refining the force developed by the actuator (Winter 90). Thus, it may be interesting to describe the strength, activation and contraction force distributions as mathematical functions of the muscle geometric space. For this purpose, we define the contraction force vector at a point P, with a function $\mathbf{f}^C$ in the form:

$$\forall P \in \mathcal{M} \quad \mathbf{f}^c(P) = -f^c(P)\,\boldsymbol{\delta}(P) \quad \text{with} \quad f^c(P) = a(P)\,s(P)\,f^M\left(l^M, v^M\right) \tag{5.41}$$

where $s(P)$ and $a(P)$ are the *strength* and *activation* distribution functions
$\boldsymbol{\delta}(P)$ is the unitary vector along the local contraction force direction
f^M is the *uniform contraction force* at maximum activation

The uniform contraction force f^M may be defined as the muscle contraction force value on the force-length-velocity diagrams as presented in Fig. 1.15. It is assumed uniform across the muscle and only depends on the muscle length and shortening velocity since it does not include the fiber direction, strength and activation. Approximately, f^M may be estimated using the force-length-velocity diagram as:

$$f^M = \frac{P^M\left(l^M, v^M\right)}{V_0^M} \tag{5.42}$$

where $P^M\left(l^M, v^M\right)$ is the force defined by the force-length-velocity diagrams
V_0^M is the initial volume of the muscle

Considering the physiological and structural properties, as well as the previous mathematical descriptions, the following statements arise:

$$\forall P \in \mathcal{M} \qquad \boldsymbol{\delta}(P) = \frac{\mathbf{d}(P)}{\|\mathbf{d}(P)\|} \qquad \text{and} \qquad f^M(l^M, v^M) = f^{\mathcal{M}} \ (\text{"}uniform\text{"}) \tag{5.43}$$

$$\forall P \in f \quad a(P) = a^f \qquad s(P) = s^f \qquad f^C(P) = a^f s^f f^M(l^M, v^M) = f^f \tag{5.44}$$

where (a^f, s^f, f^f) are constant parameters for each fiber f at current time.

Thus, the fibers may be viewed as the iso-strength, iso-activation, and iso-tension lines of the muscle, to which the contraction force vector $\mathbf{f}^C$ is always tangent, and to which the strength and activation gradient vectors are always normal:

$$\forall P \in f \qquad \mathbf{f}^C(P) = f^f \boldsymbol{\delta}(P) \qquad f^f \text{ constants a current time} \tag{5.45}$$

$$\forall P \in \mathcal{M} \qquad \nabla a(P)\, \boldsymbol{\delta}(P) = 0 \qquad \nabla s(P)\, \boldsymbol{\delta}(P) = 0 \tag{5.46}$$

These properties may be used to formulate the strength, activation, and contraction force distribution functions for muscle contraction simulation. For simplicity, in the following, we group the strength $s(P)$ and activation $a(P)$ distribution functions under the same factor $\omega(P)$ that we may name *active strength*, so that:

$$\forall P \in \mathcal{M} \qquad \mathbf{f}^c(P) = \omega(P)\, f^M(l^M, v^M)\, \boldsymbol{\delta}(P) \tag{5.47}$$

$$\text{with} \qquad \nabla\omega(P)\, \boldsymbol{\delta}(P) = 0 \qquad \omega(P) = a(P)\, s(P) \tag{5.48}$$

Fusiform. $\omega(P)$ must be a positive, decreasing function of ρ, we may suggest:

$$\frac{2a_F \rho z}{1 + a_F z^2} \frac{\partial \omega(P)}{\partial \rho} + \frac{\partial \omega(P)}{\partial z} = 0 \qquad \text{from (5.36) and (5.48), leading to:}$$

$$\omega_F(\rho, \theta, z) = \omega_F^0 + \Omega_F e^{-\left(\frac{\varpi_F y}{1+a_F z^2}\right)^{\eta_F}} \quad \text{with } \left(\omega_F^0, \Omega_F, \varpi_F, \eta_F\right) \equiv \text{constants at t} \tag{5.49}$$

Triangular. $\omega(P)$ must be a positive, decreasing function of y, we may suggest:

$$\frac{a_T y}{1+a_T z}\frac{\partial\omega(P)}{\partial y}+\frac{\partial\omega(P)}{\partial z}=0$$ from (5.38) and (5.48), leading to:

$$\omega_T(x,y,z)=\omega_T^0+\Omega_T e^{-\left(\frac{\varpi_T y}{1+a_T z}\right)^{\eta_T}} \quad \text{with } \left(\omega_T^0,\Omega_T,\varpi_T,\eta_T\right)\equiv \text{constants at t} \qquad (5.50)$$

Spiral. $\omega(P)$ must be a positive, decreasing function of ρ, and independent on z, since for spiral muscle, z is proportional to θ whichever the point:

$$\rho\tan\theta\frac{\partial\omega(P)}{\partial\rho}+\frac{\partial\omega(P)}{\partial\theta}+a_S\frac{\partial\omega(P)}{\partial z}=0$$ from (5.40) and (5.48), leading to:

$$\omega_S(\rho,\theta,z)=\omega_S^0+\Omega_S e^{-\left(\varpi_S\rho\cos\theta\right)^{\eta_S}} \quad \text{with } \left(\omega_S^0,\Omega_S,\varpi_S,\eta_S\right)\equiv \text{constants at t} \qquad (5.51)$$

Finally, for any type of muscle, time-dependency may be introduced in the active strength function $\omega(P)$ by means of factor functions $\omega_m^0(t)$, $\Omega_m(t)$, $\varpi_m(t)$, $\eta_m(t)$. Taking into account the geometric nature of the contraction force description, profit may be done of the finite element discretization in applying on the mesh nodes, contraction forces calculated from the dicretized contraction force distribution. Suggestions for applying this model to finite element simulation with other constitutive relations are presented in the next chapter (Sect. 6.3).

Conclusion

Muscle modeling has a different meaning whether the global force generated at the tendon extremity or the internal contraction force is investigated. Both approaches are necessary for complete understanding of muscle mechanics. However, there is a gap to fill in the relationship between the global force generated by the actuator and the contraction force at the sarcomeral level. Modeling this aspect would lead to more realistic simulations since the real deformation is generated by a local contraction process, accounting for the orientation of the muscle fibers as well as for their difference in strength and activation. An attempt to model this aspect has been presented in this chapter. It is based on the muscle architecture description and the uniform properties of each contractile fiber. This model must be undertaken with caution since it has not been related to any biomechanical experiment. However, it is consistent with both the local and the global aspects of the contraction phenomenon.

References

Baildon 83 R.W.A. Baildon, A.E. Chapman (1983), A new approach to the human muscle model, *J. Biomechanics*, 16, 803–809

Baratta 93 R.V. Baratta, M. Solomonow (1993), Isotonic length/force models of nine different skeletal muscles, *Medical & Biological Engineering & Computing*, 31, 449–458

Bobet 93 J. Bobet, R.B. Stein, M.N. Oguztorelli (1993), A linear time-varying model of force generation in skeletal muscle, *IEEE Trans. Biomed. Engng.*, 40, 147–58

Broek 80 J.H.M. Van den Broek, J.J.D. Van der Gon (1980), A model study of isovolumic and non-isovolumic left ventricular contractions, *J. Biomechanics,* 13, 77–87

Cutts 93 A. Cutts (1993), Muscle physiology and electromyography, in *Mechanics of Human Joints: Physiology, Pathophysiology and Treatment*, ed. by Verna Wright, Eric L. Radin. New York: Marcel Dekker

Hof 77 A.L. Hof, Jw. Van der Berg (1977), Linearity between the weighted sum of the EMGs of the human triceps surae and the total torque, *J. Biomechanics,* 10, 529–539

Hof 81a A.L. Hof, Jw. Van der Berg (1981), EMG to force processing – Part I: an electrical analogue of the Hill muscle model, *J. Biomechanics*, 14, 747–758

Hof 81b A.L. Hof, Jw. Van der Berg (1981), EMG to force processing – Part II: estimation of parameters of the Hill muscle model for the human triceps surae by means of calfergometer, *J. Biomechanics*, 14, 759–770

Hof 81c A.L. Hof, Jw. Van der Berg (1981), EMG to force processing – Part III: estimation of model parameters for the human triceps surae muscle and assessment of the accuracy by means of a torque plate, *J. Biomechanics,* 14, 771–785

Hof 81d A.L. Hof, Jw. Van der Berg (1981), EMG to force processing – Part IV: eccentric-concentric contractions on a spring-flywheel set up, *J. Biomechanics,* 14, 787–792

Hof 88 A.L. Hof (1988), Assessment of muscle force in complex movements by EMG, *International Series on Biomechanics*, Biomechanics XI-A, 110–117, Amsterdam: Free University Press

Huxley 57 H.E. Huxley (1957), The mechanism of muscular contraction, *Science,* 164, 1356–1366

Huxley 71 H.E. Huxley (1971), The structural basis of muscular contraction, *Proc. R. Soc., B.* 178, 131–149

Huxley 74 H.E. Huxley (1974), Review lecture on muscular contraction, *J. Physiol.*, 243, 1–43

Karlsson 92 D. Karlsson, B. Peterson (1992), Towards a model for force predictions in the human shoulder, *J. Biomechanics*, 25, 189–199

Kaufman 89 K.R. Kaufman, K.N. An, E.Y-S. Chao (1989), Incorporation of the muscle architecture into the muscle length tension relationship, *J. Biomechanics*, 22, 943–948

Ma 91 S. Ma, G. Zahalak (1991), A distribution moment model of energetics in skeletal muscle, *J. Biomechanics,* 24, 21–35.

Niku 89 S. Niku, J.M. Henderson (1989), Viscosity of the flexor muscles of the elbow joint under maximum contraction condition, *J. Biomechanics*, 22, 523–527

Otten 87 E. Otten (1987), A myocybernetic model of the jaw system of the rat, *Orofacial Research Group*, Bloemsingel 10, 9712 KZ Groningen, The Netherlands

Pandy 90 M.G. Pandy, F. E. Zajac, E. Sim, W.S. Levine (1990), An optimal control model for maximum-height human jumping, *J. Biomechanics*, 23, 1185–1198

Pinto 87 J.G. Pinto (1987), A constitutive description of contracting papillary muscle and its implications to the dynamics of the intact heart, *J. Biomech. Engng.,* 109, 181–191

Taber 91a L.A. Taber (1991), On a non-linear theory for muscle shells – Part I: theoretical development, *J. Biomech. Engng*., 113, 56–62

Taber 91b L.A. Taber (1991), On a non-linear theory for muscle shells – Part II: application to the beating left ventricle, *J. Biomech. Engng*., 113, 63–71

Winter 90 D.A. Winter (1990), Muscle mechanics, in *Biomechanics and Motor Control of the Human Movements,* 2nd edn. New York: John Wiley & Sons

Wong 71 A.Y.K. Wong (1971), Mechanics of cardiac muscle, based on Huxley's model: mathematical simulation of isometric contraction, *J. Biomechanics,* 4, 529–540

Wood 81 J.E. Wood, R.W. Mann (1981), A sliding filament cross-bridge ensemble model of muscle contraction for mechanical transients, *Math. Biosc.,* 57, 211–263

Zahalak 90 G.I. Zahalak, S.-P. Ma (1990), Muscle activation and contraction: constitutive relations based directly on cross-bridges kinetics, *J. Biomech. Engng.,* 112, 52–62

Zajac 86 F.E. Zajac, E.L. Topp, P.J. Stevenson (1986), A dimensionless musculotendon model, *Proc. 8th A. Conf. IEEE Engng. Med. Biol. Soc.,* 601–604, Dallas-Ft Worth, TX, 7–10 Nov.

Zajac 89 F.E. Zajac (1989), Muscle and tendon: properties, models scaling and application to biomechanics and motor control, *CRC Critic. Rev. in Biomed. Engng.*, 17, 359–411

6 Application Perspectives

Various approaches have been followed for soft tissues modeling in computer graphics. Most of them rely on the assumptions of geometrical and physical linearity of the material. However, applying biomechanical constitutive relations is likely to produce more realistic simulations. Given their empirical nature, these relations may not fit exactly with the general theoretical relations and resolution methods. This leads therefore to various individual approaches for adjusting the models to the technical requirements of the software planned for the simulation. In this final chapter, some directions and ideas are suggested for matching biomechanical models such as those presented in Chapt. 4, with theoretical relations and methods as presented in Chapts. 2 and 3. It does not claim to select "the" model that must be used for soft tissue simulation. It rather offers a synthesis of approaches and methods, and makes suggestions for applying the available models to soft tissue simulation.

6.1 Physically-Based Modeling

6.1.1 General Approach

State Variables. The main concepts involved in mechanics are the notions of action and motion. In particular, the mechanical state of a system may be described in terms of stress and strain variables. These are (Sect. 2.1.2):

$\mathbf{e}$, $\boldsymbol{\sigma}$: Eulerian strain, Cauchy stress tensors in Eulerian configuration

$\mathbf{E}$, $\mathbf{S}$: Lagrange strain, Kirchhoff stress tensors in Lagrangian configuration

$\boldsymbol{\epsilon}$, $\boldsymbol{\sigma}$: linearized strain, Cauchy stress tensors for infinitesimal strains

Equation of motion. The dynamics of a system may thus be fully described with an equation of motion relating its state variables to the external actions applied onto it. Motion may generally be described using the *virtual work* principle:

$$\delta\hat{W}^{acc} = \delta\hat{W}^{int} + \delta\hat{W}^{ext} \tag{6.1}$$

where	$\delta\hat{W}^{acc}$	inertial virtual work due to the acceleration
	$\delta\hat{W}^{int}$	internal virtual work due to the stresses and strains
	$\delta\hat{W}^{ext}$	external virtual work due to the external actions

which takes different forms whether this principle is formulated in the Eulerian (2.48) or the Lagrangian (2.52) configuration, or under the infinitesimal strain condition (2.55). Equation (6.1) may then be solved in order to determine the state of the system resulting from the external actions applied onto it.

Constitutive Relationships. In (6.1), both state variables $(\mathbf{e},\boldsymbol{\sigma})$, $(\mathbf{E},\mathbf{S})$, $(\boldsymbol{\epsilon},\boldsymbol{\sigma})$ are involved, so it cannot be solved without considering the constitutive relationship of the material for relating the stress to the strain. Various theoretical and empirical constitutive relationships have been presented in Sects. 2.2 and 2.3 and Chapt. 4. Common assumptions are linear constitutive relationships such as:

Hooke:	$\boldsymbol{\sigma} = \mathbf{K}^{E}\boldsymbol{\epsilon}$	for linear elasticity	(2.92)	(6.2)
Kelvin–Voigt:	$\boldsymbol{\sigma} = \mathbf{K}^{E}\boldsymbol{\epsilon} + \mathbf{D}^{V}\dot{\boldsymbol{\epsilon}}$	for linear viscoelasticity	(2.103)	(6.3)

Complemented with a constitutive relationship, (6.1) may be solved for the remaining state variable using appropriate resolution methods.

Geometric Discretization. In the general case, geometric discretization of the system is necessary in order to approximate the continuous problem by a finite number of equations defined for the mesh nodes. Various discretization methods may be applied for this purpose. All, however, derive from the most general *finite element* method presented in Sect. 3.1. For a general non-linear viscoelastic material, geometric discretization of (6.1) leads to the tensorial form:

$$\mathbf{M}\,\ddot{\mathbf{U}} + \boldsymbol{\Pi}\left(\mathbf{U},\dot{\mathbf{U}}\right) = \mathbf{L} \qquad (3.14) \tag{6.4}$$

where $\mathbf{M}$ is the nodal mass matrix, $\mathbf{U}$, $\boldsymbol{\Pi}$, $\mathbf{L}$, the nodal displacement, internal and external force vectors. The solution of (6.4) is thus obtained in terms of the mesh node coordinates which approximate the continuous deformable system.

Temporal Integration. In the general case, due to the non-linearities involved in the problem, no analytical solution defining the system behavior as a continuous function of time may be carried out. It is therefore necessary to convert the time-continuous equation of motion and constitutive relation into incremental forms, so that within each time interval the problem may be handled as linear (Sect. 3.2). As a result, the time-continuous problem is replaced by a finite system of linear tensorial equations defined over respective intervals, of the form:

$$\mathbf{M}_n \Delta\ddot{\mathbf{U}} + \mathbf{D}_n \Delta\dot{\mathbf{U}} + \mathbf{K}_n \Delta\mathbf{U} = \Delta\mathbf{L} \qquad \text{over } [t_n, t_{n+1}] \tag{6.5}$$

with $\Delta\mathbf{U} = \mathbf{U}_{n+1} - \mathbf{U}_n$ nodal displacement increment vector
$\Delta\mathbf{L} = \mathbf{L}_{n+1} - \mathbf{L}_n$ nodal external force increment vector
$\mathbf{M}_n$, $\mathbf{D}_n$, $\mathbf{K}_n$ tangent mass, damping, stiffness matrix

Several incremental methods exist, such as those briefly presented in Sect. 3.3, which allow the incremental simulation of the system behavior. Various applications of this methodology to soft tissue simulation are presented in the following section.

6.1.2 Lagrange Formalism-Based Approaches

Physically-based modeling is a common approach in computer animation for natural phenomenon simulation. A pioneering physically-based approach for realistic deformation simulation has been introduced by Terzopoulos et al. (Terzopoulos 87). This approach consists in applying the *Lagrange formalism*, the finite difference geometric discretization method and a *semi-implicit* temporal integration method to simulate the dynamics of deformable objects.

Lagrange Formalism. Using generalized coordinates q_i ($\mathbf{x} = \mathbf{x}(q_i, t)$), and assuming a constant mass density ($\rho \equiv \rho_0$), the equation of motion (6.1) of the continuous medium Ω may be converted into the *Lagrange* equation as:

$$\frac{d}{dt}\left(\frac{\partial C}{\partial \dot{q}_i}\right) - \frac{\partial C}{\partial q_i} = f^{int}_{q_i} + f^{ext}_{q_i} \qquad \text{with} \qquad C = \frac{1}{2}\int_V \rho \left(\frac{d\mathbf{x}}{dt}\right)^2 dV \tag{6.6}$$

where C is the kinetic energy of Ω
$f^{int}_{q_i}$ projections on q_i of the internal force vector
$f^{ext}_{q_i}$ projections on q_i of the external force vector

In this approach, Terzopoulos et al. considered that the internal force vector $\mathbf{f}^{int}$ derives from internal viscous and elastic potential energies as:

$$f_{q_i}^{int} = -\frac{\partial W_{visco}^{int}}{\partial \dot{q}_i} - \frac{\partial W_{elast}^{int}}{\partial q_i} \quad \text{with} \quad W_{visco}^{int} = \frac{1}{2}\eta\,\dot{q}_i^{\,2} \tag{6.7}$$

where W_{elast}^{int} is the strain energy function
W_{visco}^{int} is the viscous energy function assumed linear
η is the constant uniform damping density

and proposed a strain energy function as:

$$W_{elast}^{int} = \int_L (\alpha.\text{stretching} + \beta.\text{bending} + \gamma.\text{twisting})dl \quad \text{for a curve} \tag{6.8}$$

$$W_{elast}^{int} = \int_S (\alpha.\text{stretching} + \beta.\text{bending})ds \quad \text{for a surface} \tag{6.9}$$

$$W_{elast}^{int} = \int_V \left\| \mathbf{C} - \mathbf{C}^0 \right\|_\alpha^2 dV \quad \text{for a volume} \tag{6.10}$$

where $\mathbf{C}$ is the Cauchy–Green right dilation tensor of the deformation
$\mathbf{C}^0$ is the dilation tensor associated to the natural shape of the body
$\| \ \|_\alpha$ is α-weighted matrix norm, and α, β, γ tuning constants

Using (6.7), and assuming also a constant uniform mass density ρ, the Lagrange equation of motion (6.6) was finally obtained in the form:

$$\frac{d}{dt}\left(\rho\frac{dq_i}{dt}\right) + \eta\frac{dq_i}{dt} + \frac{\partial W_{elast}^{int}}{\partial q_i} = f_{q_i}^{ext} \tag{6.11}$$

Finite Difference Discretization. The finite difference method is based on spatial subdivisions of Ω along its dimensions. Continuous functions applied onto Ω may then be approximated to the first order by differentiation between the nodal values. This method, in fact, corresponds to the finite element method, applied with linear interpolation functions and a grid-subdivision of Ω. For example, for a deformable surface, using subdivision steps $h_1 \times h_2$, the first and second derivative of the 2D grid function $u_{m,n} = u[m,n]$ is defined as (Terzopoulos 87):

central: $D_{11}(u)[m,n] = D_1^-[D_1^+(u)]\,[m,n] \quad D_{22}(u)[m,n] = D_2^-[D_2^+(u)]\,[m,n]$ (6.12) *forward*: $$D_1^+(u)[m,n]=\frac{u_{m+1,n}-u_{m,n}}{h_1} \qquad D_2^+(u)[m,n]=\frac{u_{m,n+1}-u_{m,n}}{h_2} \tag{6.13}$$

$$D_{12}^+(u)[m,n]=D_{21}^+(u)[m,n]=D_1^+\left[D_2^+(u)\right][m,n]$$

backward: $$D_1^-(u)[m,n]=\frac{u_{m,n}-u_{m-1,n}}{h_1} \qquad D_2^-(u)[m,n]=\frac{u_{m,n}-u_{m,n-1}}{h_2} \tag{6.14}$$

$$D_{12}^-(u)[m,n]=D_{21}^-(u)[m,n]=D_1^-\left[D_2^-(u)\right][m,n]$$

As a result of this discretization, the equation of motion (6.11) takes the form:

$$\mathbf{M}\,\ddot{\mathbf{U}}+\mathbf{D}\,\dot{\mathbf{U}}+\mathbf{K}(\mathbf{U})\,\mathbf{U}=\mathbf{L} \tag{6.15}$$

with	**M** and **D**	mass and damping diagonal matrix
	K	deformation dependent stiffness matrix
	U, **L**	nodal displacement and external nodal force vectors

Semi-Implicit Integration. For temporal integration, Terzopoulos et al. proposed to convert the non-linear ordinary differential system (6.14) into a sequence of linear algebraic systems by means of a finite difference time-discretization as:

$$\ddot{\mathbf{U}}_n=\frac{\mathbf{U}_{n+1}-2\mathbf{U}_n+\mathbf{U}_{n-1}}{\Delta t^2} \qquad \dot{\mathbf{U}}_n=\frac{\mathbf{U}_{n+1}-\mathbf{U}_{n-1}}{2\Delta t} \qquad \mathbf{U}_n=\mathbf{U}(t_n) \tag{6.16}$$

and to apply a semi-implicit procedure (Sect. 3.3.2) to integrate them through time, as:

$$\mathbf{A}_n\mathbf{U}_{n+1}=\mathbf{L}_{n+1}-\mathbf{Y}_n \qquad \text{where} \qquad \Delta t \text{ is the time step} \tag{6.17}$$

$$\mathbf{A}_n=\mathbf{K}_n+\frac{1}{2\Delta t}\mathbf{D}+\frac{1}{\Delta t^2}\mathbf{M} \qquad \mathbf{M},\ \mathbf{D}\ \text{constant mass, damping matrix}$$

$$\mathbf{Y}_n=-\left(\frac{1}{\Delta t^2}\mathbf{M}+\frac{1}{2\Delta t}\mathbf{D}\right)\mathbf{U}_n-\left(\frac{1}{\Delta t}\mathbf{M}-\frac{1}{2}\mathbf{D}\right)\mathbf{V}_n \qquad \mathbf{V}_n=\frac{\mathbf{U}_n-\mathbf{U}_{n-1}}{\Delta t}$$

The model was later extended to inelastic behavior such as viscoelasticity, plasticity and fracture by means of a general strain energy function in the form of control continuity generalized spline kernels (Terzopoulos 88).

6.1.3 Linear Finite Element Approaches

Several models for skin and muscle have been proposed following the linear finite element approach. Larrabee et al. applied the finite element method to simulate the effect of skin flap design. Skin was modeled as a linear elastic membrane with regularly spaced nodes, connected by linear springs to subcutaneous attachments (Larrabee 86). Gourret et al. applied the finite element method to grasping simulation including deformation of the grasped objects and of the skin of the fingers (Fig. 6.1) (Gourret 89). Essa et al. proposed to use the linear finite element mode superposition method based on Sect. 3.3.4 to simulate the dynamics of deformable objects. In this approach, they considered the deformable body as a linear viscoelastic isotropic incompressible Kelvin–Voigt material as:

$$\boldsymbol{\sigma} = \mathbf{K}^{E}\boldsymbol{\epsilon} + \mathbf{D}^{V}\dot{\boldsymbol{\epsilon}} \qquad (2.103) \qquad \text{with } \mathbf{K}^{E},\ \mathbf{D}^{V} \text{ constant matrix} \qquad (6.18)$$

with an elastic component defined by a constant elastic stiffness matrix $\mathbf{K}^{E}$ in terms of the *Young* modulus E and *Poisson* coefficient ν (2.94), and with a material damping matrix $\mathbf{D}^{V}$ defined at the discretization stage following to the *Rayleigh damping* principle. The discrete equation of motion was obtained as:

$$\mathbf{M}\,\ddot{\mathbf{U}} + \mathbf{D}\,\dot{\mathbf{U}} + \mathbf{K}\,\mathbf{U} = \mathbf{L} \qquad (3.19) \qquad \text{with } \mathbf{M},\ \mathbf{D},\ \mathbf{K} \text{ constant matrix} \qquad (6.19)$$

in which, according to the *Rayleigh damping* option, $\mathbf{D}$ is a linear combination of the mass and stiffness matrix $\mathbf{M}$ and $\mathbf{K}$, as:

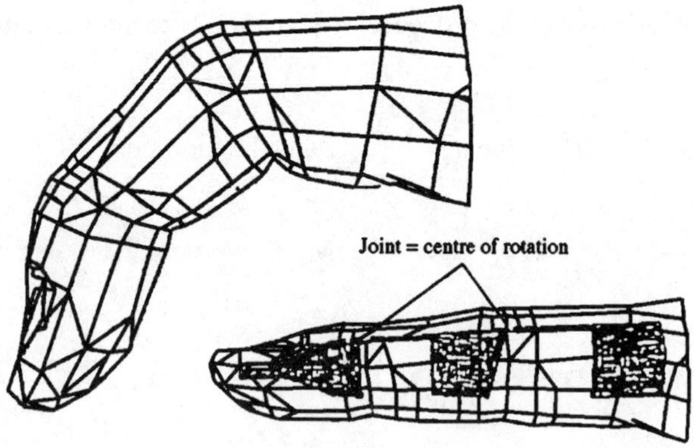

Fig. 6.1. Finite element model of finger
(reprinted from (Gourret 91) with permission from Elsevier Science)

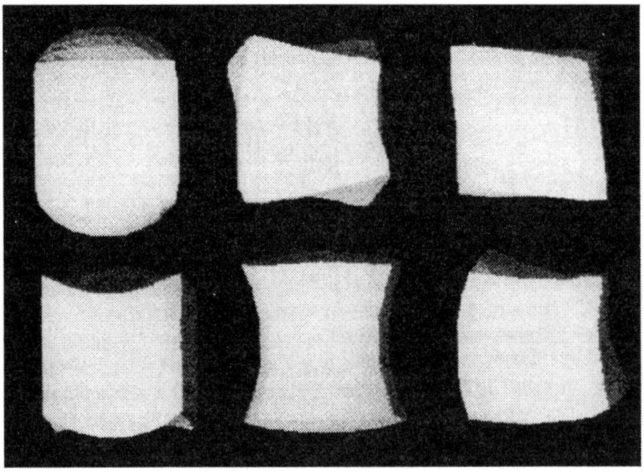

Fig. 6.2. Deformation modes of a 27-node finite element cube
(reprinted from (Essa 93) with permission from the authors)

$\mathbf{D} = \alpha\,\mathbf{M} + \beta\,\mathbf{K}$ where α, β are parameters for controlling system stability

Using the mode superposition method (Sect. 3.3.4), the deformation of a solid may be represented as a linear combination of its deformation modes (Fig. 6.2). The modal equation of motion was then obtained in the form:

$$\ddot{\mathbf{V}} + \boldsymbol{\Lambda}\,\dot{\mathbf{V}} + \boldsymbol{\Omega}^2\,\mathbf{V} = \tilde{\mathbf{L}} \qquad \text{with} \qquad (6.20)$$

$$\tilde{\mathbf{M}} = \boldsymbol{\Psi}^{\mathrm{T}}\mathbf{M}\boldsymbol{\Psi} \equiv \mathbf{I} \qquad \tilde{\mathbf{D}} = \boldsymbol{\Psi}^{\mathrm{T}}\mathbf{D}\boldsymbol{\Psi} = \boldsymbol{\Lambda} \qquad \tilde{\mathbf{K}} = \boldsymbol{\Psi}^{\mathrm{T}}\mathbf{K}\boldsymbol{\Psi} = \boldsymbol{\Omega}^2 \qquad \tilde{\mathbf{L}} = \boldsymbol{\Psi}^{\mathrm{T}}\mathbf{L}$$

which may be solved using common algorithms for linear algebraic system.

Following this approach, Chen et al. developed a linear finite element model of skeletal muscle for computer animation, including a contraction process based on Zajac's muscle force model (Sect. 5.2.2). For simplification, the finite element analysis was performed on a prismatic bounding box embedding the muscle object (Fig. 6.3), and the resulting deformations were mapped onto the muscle using the free-form deformation principle (FFD) introduced by Sederberg and Parry (Sederberg 86). Assuming the contraction forces greater than the passive response in the physiological range of functioning, Chen et al. neglected the passive non-linearities, and considered muscle as a homogeneous, incompressible, linear, isotropic, viscoelastic material as (6.19). Muscle contraction was then simulated by incrementally applying biomechanically-based loads on the mesh (Chen 92).

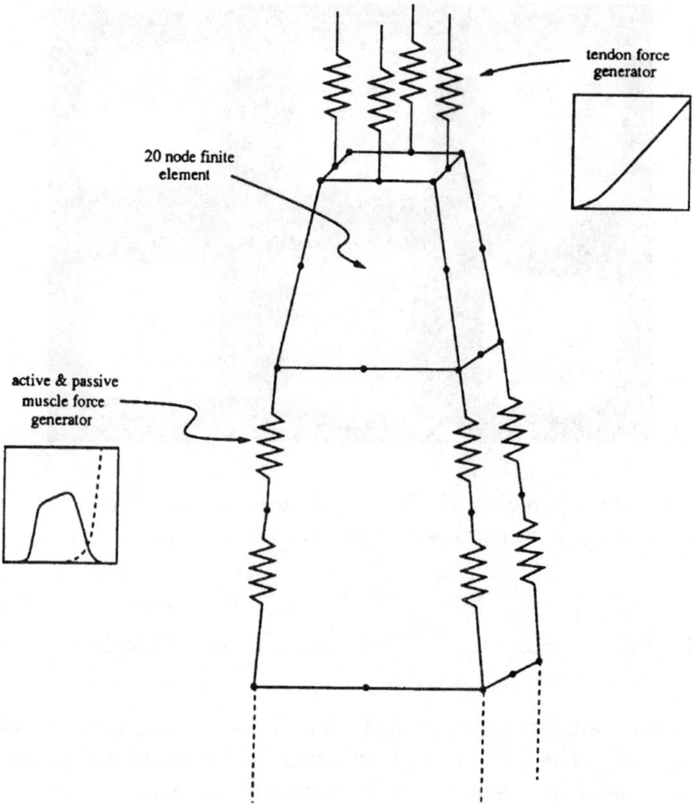

Fig. 6.3. Chen's finite element mesh and contraction model of muscle (reprinted from (Chen 92) with permission of ACM Siggraph)

Recently, Bro-Nielsen et al. proposed a linear finite element modeling approach, including common computational improvements such as matrix condensation techniques, in order to achieve real-time surgery simulation (Bro-Nielsen 96).

6.1.4 Specific Analyses

Other approaches have been proposed for soft tissue modeling in computer animation. Platt and Badler developed a facial animation model, in which skin was represented as a warped tension net, with nodes connected together by Hookean springs, and muscles as macroscopic fibers connecting the skin nodes to the underlying bone. Facial expressions were thus simulated by applying forces on the skin mesh by means of the muscle fibers (Fig. 6.4) (Platt 81).

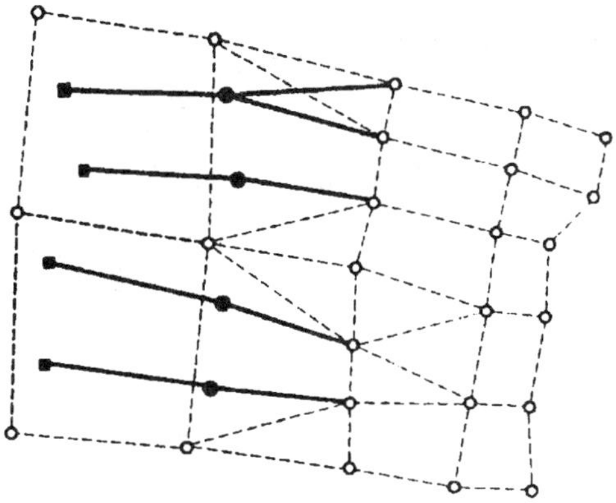

Fig. 6.4. Muscles and skin tension net
(reprinted from (Platt 81) with permission of ACM Siggraph)

Waters et al. improved this approach in designing muscle vectors following the major direction and insertion of the real facial muscle fibers. Three types of muscles were modeled: linear, sphincter, and sheet. The linear and sheet muscle contractions were designed as vectors, while the sphincter muscle contraction was designed as elliptical (Waters 87). Gascuel and Verroust also proposed a model of skin with spring-like muscles, in which the skin layer was represented by a B-spline surface attached by its control points (Gascuel 91). Terzopoulos and Waters modeled facial tissues using a trilayer spring lattice structure, each layer representing respectively the skin, the subcutaneous tissue, and the underlying muscle layers, with different stiffness parameters according to the relevant tissue (Fig. 6.5) (Terzopoulos 90–91).

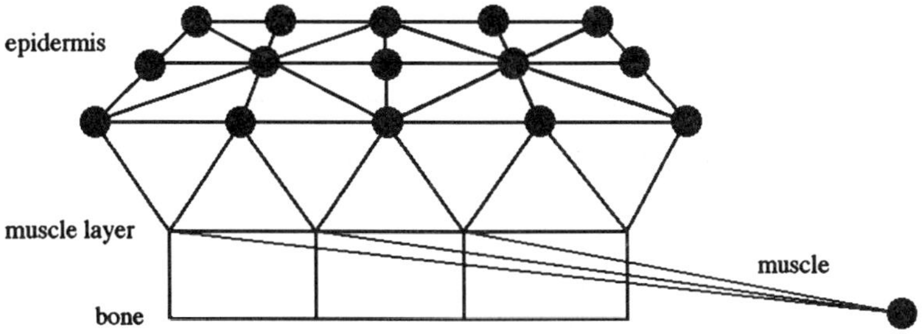

Fig. 6.5. Trilayer lattice
(reprinted from (Terzopoulos 90) with permission of John Wiley & Sons)

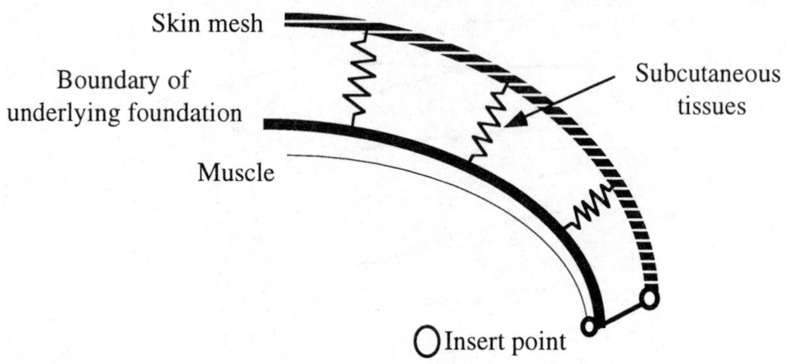

Fig. 6.6. Model for connections between skin and other body tissues (reprinted from (Wu 95) with permission of John Wiley & Sons)

Pieper et al. used a linear model of skin to perform simple facial surgical simulation. In their approach, both the skull and skin surfaces extracted from CT images were mapped into cylindrical coordinates, and a home-grown static displacement-based method was used to solve the elastic equilibrium equation (Pieper 92). Wu et al. modeled skin as a thin membrane connected to strong spring-like muscles by means of soft bilinear springs representing the respective contributions of elastin and collagen fibers to the elastic behavior of the subcutaneous tissues (Wu 95). The connection forces f_{sb}, f_m were expressed as:

$$f_{sb} = -k_{sb}\left[\, \left\| x_{sk}^t - x_m^t \right\| - \left\| x_{sk}^0 - x_m^0 \right\| \,\right] \frac{x_{sk}^t - x_m^t}{\left\| x_{sk}^t - x_m^t \right\|} \text{ from the subcutaneous tissues} \quad (6.21)$$

$$f_m = -k_m\left[\, \left\| x_{sk}^t - x_m^t \right\| - \left\| x_{sk}^0 - x_m^0 \right\| \,\right] \frac{x_{sk}^t - x_m^t}{\left\| x_{sk}^t - x_m^t \right\|} \text{ from the muscles} \quad (6.22)$$

where k_m, k_{sb} are the muscle and subcutaneous tissue spring stiffnesses
x_m^t, x_{sk}^t are the attachment node positions on muscle and skin at time t

Recently, Koch et al. applied the general approach for finite element curve and surface free-form shape designing, proposed by Celniker and Gossard (Celniker 91), to surgical planning and prediction of human facial shape after craniofacial and maxillofacial surgery (Koch 96). This approach follows Terzopoulos' approach presented in Sect. 6.1.2, but uses finite element primitives instead of a finite difference discretization, and applies a strain energy function of the form:

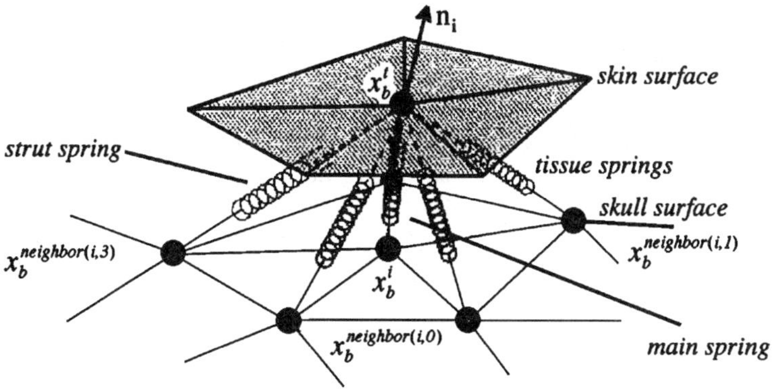

Fig. 6.7. Spring mesh between skin and skull
(reprinted from (Koch 96) with permission of ACM Siggraph)

$$W^{int} = \int_{\Omega} (\alpha \text{ stretching} + \beta \text{ bending}) \, d\Omega \qquad (6.9) \qquad (6.23)$$

where α and β are user-defined parameters for smoothness tuning. Interactive control of the shape is then achieved by parametrizing the shape with sculpting loads or geometric constraints (Celniker 91). In Koch's application, the facial surface and inner structures (bones or muscles) were obtained from the patient's CT or MRI data, and the effect of the subcutaneous tissues was simulated by connecting the finite element skin surface nodes with nodal springs to the skull as shown in Fig. 6.7 (Koch 96).

6.2 Suggestions for Biomechanical Finite Element Simulation

6.2.1 Non-linear Finite Element Models

Numerous non-linear finite element models have been developed for soft-tissue simulation, mainly for lungs (Liu 78, Pao 78, Vawter 80, Karakaplan 80, Lee 83) and ventricles (Janz 73, Janz 74, Christie 80-82, Needleman 83, Huang 90, Huygue 91, Taber 91a-b). An incremental dynamic analysis has also been presented by Deng et al. who developed a non-linear thick shell model including a skin layer, a sliding layer, and a muscle layer, to simulate the closure of skin excisions on facial tissue (Deng 88). Usually, the material tangent stiffness matrix $\mathbf{K}_n^M$ of the incremental constitutive relationship (3.40) is obtained by second-order derivation of a strain energy function W with respect to the strain tensor $\mathbf{E}$:

$$\Delta \mathbf{S} = \mathbf{K}_n^M \Delta \mathbf{E} + \mathbf{S}_n^R \qquad \text{with} \qquad \mathbf{K}_n^M = \frac{\partial^2 W}{\partial \mathbf{E}_n^2} \tag{6.24}$$

Horowitz et al. presented an extended development of such incremental formulation as they led for 2D (Horowitz 88a) and 3D (Horowitz 88b) finite element heart muscle modeling with the fibrous constitutive model presented in Sect. 4.3.3 (Horowitz 88c). For an incompressible ventricular wall segment, the Kirchhoff stress $\mathbf{S}$ was derived from a pseudoelastic strain energy function W based on Lanir's suggestion (Lanir 83):

$$W = W_1(I_1, I_2) + W_2(I_3) \qquad \text{with} \tag{6.25}$$

$$W_1 = \sum_k W_k = \sum_k S_k \int_\Omega R_k(\mathbf{u}) w_k^*(\gamma_n') \, d\Omega \qquad W_2 = L(I_3 - 1)$$

$$\mathbf{S} = \frac{\partial W_1}{\partial \mathbf{E}} + L \frac{\partial I_3}{\partial \mathbf{E}} \tag{6.26}$$

where I_3 is the third invariant the Cauchy–Green right dilation tensor $\mathbf{C}$, and L is a Lagrange multiplier accounting for the hydrostatic pressure of the incompressible embedding matrix. The stress and stress increment at n were then derived from (6.26) as:

$$\mathbf{S}_n = \frac{\partial W_1}{\partial \mathbf{E}_n} + L_n \frac{\partial I_3}{\partial \mathbf{E}_n} \qquad (L \equiv L_n, \text{ since pressure may vary with time) i.e.:} \tag{6.27}$$

$$\mathbf{S}_n = S_c \int_{-\frac{\pi}{3}}^{\frac{\pi}{3}} \int_0^{2\pi} \int_0^{\pi} f_c^*(\varepsilon_n') R_c(\alpha, \theta, \phi) \frac{\partial \varepsilon_n'}{\partial \mathbf{E}_n} J \, d\phi d\theta d\alpha + L_n \frac{\partial I_3}{\partial \mathbf{E}_n}$$

$$\Delta \mathbf{S} = \frac{\partial \mathbf{S}_n}{\partial \mathbf{E}_n} \Delta \mathbf{E} + \frac{\partial \mathbf{S}_n}{\partial L_n} \Delta L = \left(\frac{\partial^2 W_1}{\partial \mathbf{E}_n^2} + L_n \frac{\partial^2 I_3}{\partial \mathbf{E}_n^2} \right) \Delta \mathbf{E} + \frac{\partial I_3}{\partial \mathbf{E}_n} \Delta L \qquad \text{i.e.:} \tag{6.28}$$

$$\Delta \mathbf{S} = \left(S_c \int_{-\frac{\pi}{3}}^{\frac{\pi}{3}} \int_0^{2\pi} \int_0^{\pi} f_c^*(\varepsilon_n') R_c(\alpha, \theta, \phi) \frac{\partial^2 \varepsilon_n'}{\partial \mathbf{E}_n^2} J \, d\phi d\theta d\alpha + L_n \frac{\partial^2 I_3}{\partial \mathbf{E}_n^2} \right) \Delta \mathbf{E} + \frac{\partial I_3}{\partial \mathbf{E}_n} \Delta L$$

As a result, the tangent elastic stiffness matrix $\mathbf{K}_n^M$ required for the incremental stress-strain formulation was obtained in the form:

$$\mathbf{K}_n^M = \frac{\partial^2 W}{\partial \mathbf{E}_n^2} = \frac{\partial^2 W_1}{\partial \mathbf{E}_n^2} + L_n \frac{\partial^2 I_3}{\partial \mathbf{E}_n^2} \qquad \text{with W as defined in (6.25), i.e.:} \tag{6.29}$$

$$\mathbf{K}_n^M = S_c \int_{-\frac{\pi}{3}}^{\frac{\pi}{3}} \int_0^{2\pi} \int_0^{\pi} \left[\int_0^{\varepsilon_c} \frac{D_c(x)}{1+2x} dx \right] R_c(\alpha, \theta, \phi) \frac{\partial^2 \varepsilon_n'}{\partial \mathbf{E}_n^2} J \, d\phi d\theta d\alpha + L_n \frac{\partial^2 I_3}{\partial \mathbf{E}_n^2} \tag{6.30}$$

The resolution of the equation of motion was then performed using a finite element incremental-iterative modified Newton–Raphson scheme (Sect. 3.3.3), in which the tangent stiffness matrix $\mathbf{K}_n$ (3.88) was updated after each load increment only. The approach was applied to finite element modeling of canine myocardial strips (Horowitz 88ab).

6.2.2 Incremental Constitutive Modeling

As shown above, given their empirical nature, the available biomechanical constitutive relationships are likely not to fit exactly with the general theoretical incremental forms described in Sect. 3.2. Therefore, the important point is not to rigorously obtain forms identical to the theoretical relations, but to find similar ones, compatible with the practical aspects of the resolution methods, and respecting the meaning of the original constitutive relationship.

Linear Viscoelasticity. The incremental formulation for linear viscoelasticity was described in Sect. 3.2.5 as:

$$\Delta \mathbf{S} = \left[\mathbf{K}^E - \mathbf{K}^V \right] \Delta \mathbf{E} - \mathbf{S}_n^V \qquad \text{or} \qquad \Delta \boldsymbol{\sigma} = \left[\mathbf{K}^E - \mathbf{K}^V \right] \Delta \boldsymbol{\epsilon} - \boldsymbol{\sigma}_n^V \tag{6.31}$$

All relations presented in Sect. 4.2.2 are candidates to such a formulation. Examples are given hereafter for a *quasi-static evolution*. The Lagrangian stress defined by:

(4.49) (Hirsch 68) leads to

$$\Delta T = b\Delta\varepsilon + (c\varepsilon - aT)\Delta t \tag{6.32}$$

(4.53) (Galford 70) leads to

$$\Delta T = \frac{q_1}{p_1} \Delta\varepsilon - \frac{T}{p_1} \Delta t \tag{6.33}$$

(4.59) (Sanjeevi 82) leads to

$$\Delta T = (E_1 + 2E_2\varepsilon)\Delta\varepsilon + \eta_2 \left(\frac{d\varepsilon}{dt} \right)^2 \tag{6.34}$$

Considering integral forms such as Fung's QLV relation defined by (4.10), a series decomposition as presented in Sect. 2.3.4 may be developed using the relaxation times τ_1, τ_2 of the material, and an incremental relation may be developed in a form similar to (3.50) as:

$$\Delta \mathbf{T} = \Delta \mathbf{T}^{(e)} - \mathbf{D}^{P} \sum_{i=1}^{n_P} P_i \, \mathbf{P}_i^{(e)} - \mathbf{D}^{Q} \sum_{i=1}^{n_Q} Q_i \, \mathbf{Q}_i^{(e)} \qquad \text{with} \tag{6.35}$$

$$\mathbf{P}_i^{(e)} = \int_0^t \frac{\mathbf{T}_n^{(e)}(\tau)}{\beta_i} e^{-\frac{t-\tau}{\beta_i}} d\tau \qquad \text{and} \qquad \mathbf{Q}_i^{(e)} = \int_0^t \frac{\mathbf{T}_n^{(e)}(t)}{\gamma_i} e^{-\frac{t-\tau}{\gamma_i}} d\tau \tag{6.36}$$

where $\mathbf{T}_n^{(e)}$ is the Lagrange elastic stress response as defined by (4.12).

Non-linear Elasticity. All the uniaxial elastic constitutive relations presented in Sect. 4.1 and 4.2 may be used to derive elastic the tangent modulus, for example:

(4.2) (Wertheim 47) leads to
$$\Delta T = \frac{2\varepsilon}{b + 2aT} \Delta\varepsilon \tag{6.37}$$

(4.20) (Feit 79) leads to
$$\Delta T = abe^{b\varepsilon} \Delta\varepsilon \tag{6.38}$$

(4.21) (Haut 69) leads to
$$\Delta T = ae^{b\varepsilon}\left[1 + (b-2)\varepsilon - b\varepsilon^2\right]\Delta\varepsilon \tag{6.39}$$

(4.33) (Diamant 72) leads to
$$\Delta T = \left(\frac{1}{E} + \frac{\gamma}{2T}\sqrt{\frac{E}{T}}\right)^{-1} \Delta\varepsilon \tag{6.40}$$

(4.43) (Kwan 89)
$$\Delta T = \begin{bmatrix} E_S \\ \left[E_S + \gamma_1(E_e - E_S)\right] \\ \left[E_S + (\gamma_1 + \gamma_2)(E_e - E_S)\right] \\ E_e \\ (1-\beta_1)E_e \\ \beta_3 E_e \end{bmatrix} \Delta\varepsilon \ \text{ for } \begin{cases} 0 < \varepsilon \le \varepsilon_{S1} \\ \varepsilon_{S1} < \varepsilon \le \varepsilon_{S2} \\ \varepsilon_{S2} < \varepsilon \le \varepsilon_{S3} \\ \varepsilon_{S3} < \varepsilon \le \varepsilon_{U1} \\ \varepsilon_{U1} < \varepsilon \le \varepsilon_{U2} \\ \varepsilon_{U2} < \varepsilon \le \varepsilon_{U3} \end{cases} \tag{6.41}$$

Concerning 3D modeling, the strain energy functions defined in Sect. 4.3 may be used to formulate a tangent stiffness matrix following the approach described in Sect. 6.2.1, as for example:

$$\Delta \mathbf{S} = \mathbf{K}_n^M \Delta \mathbf{E} \qquad \mathbf{K}_n^M = \frac{\partial^2 W}{\partial \mathbf{E}_n^2} \tag{6.42}$$

(4.70) (Klosner 69)
$$\begin{cases} K_{11}^M = 32\,(E_{22}+E_{33}+1)^2\,[C_3+3C_4(I_2-3)] \\ K_{22}^M = 32\,(E_{11}+E_{33}+1)^2\,[C_3+3C_4(I_2-3)] \\ K_{33}^M = 32\,(E_{11}+E_{22}+1)^2\,[C_3+3C_4(I_2-3)] \end{cases} \tag{6.43}$$

(4.74) (Allaire 77)
$$\begin{cases} K_{11}^M = 8\,C_3\,(2E_{22}+1)^2\,(2E_{33}+1)^2 \\ K_{22}^M = 8\,C_3\,(2E_{11}+1)^2\,(2E_{33}+1)^2 \\ K_{33}^M = 8\,C_3\,(2E_{11}+1)^2\,(2E_{22}+1)^2 \end{cases} \tag{6.44}$$

(4.86) (Vawter 80)
$$\begin{cases} K_{11}^M = \frac{C}{2D_A} e^{(\alpha I_1^2+\beta I_2)}\left[[4\alpha I_1 + 2\beta(E_{22}+E_{33}+1)]^2 + 8\alpha\right] \\ K_{22}^M = \frac{C}{2D_A} e^{(\alpha I_1^2+\beta I_2)}\left[[4\alpha I_1 + 2\beta(E_{11}+E_{33}+1)]^2 + 8\alpha\right] \\ K_{33}^M = \frac{C}{2D_A} e^{(\alpha I_1^2+\beta I_2)}\left[[4\alpha I_1 + 2\beta(E_{11}+E_{22}+1)]^2 + 8\alpha\right] \end{cases} \tag{6.45}$$

6.2.3 Tendon, Skin and Passive Muscle Modeling

General Approach. The analyses provided in Chapts. 1 and 4 of the mechanical properties and models of soft tissues have led to a characterization of their behavior as quasi-linear viscoelastic (QLV) as outlined by Fung (Fung 72) and described in Sect. 4.1.2. According to Fung, this formulation may be extended to 3D modeling with a relationship of the form (Fung 93):

$$\mathbf{S}(t) = \mathbf{G}(t)\,\mathbf{S}^{(e)}(0) + \int_{-\infty}^{t} \mathbf{G}(t-\tau)\frac{\partial \mathbf{S}^{(e)}[\mathbf{E}(\tau)]}{\partial \tau}\,d\tau \tag{6.46}$$

where $\mathbf{S}(t)$ is the second Piola–Kirchhof stress tensor,
$\mathbf{S}^{(e)}[\mathbf{E}]$ the pure elastic response of the material
and $\mathbf{G}(t)$ the tensorial reduced relaxation function.

Considering the lack of experimental data to fully determine the terms of the tensorial reduced relaxation function $\mathbf{G}(t)$, we suggest assuming $\mathbf{G}(t) \equiv G(t)$ as a scalar function, and to consider the form as suggested by Fung in (4.28) as:

$$G(t) = \alpha \ln(t) + \beta \qquad \alpha = -\frac{c}{1 + c\ln\left(\frac{\tau_2}{\tau_1}\right)} \qquad \beta = \frac{1 - c\,\gamma + c\ln(\tau_2)}{1 + c\ln\left(\frac{\tau_2}{\tau_1}\right)} \tag{6.47}$$

Concerning the elastic response, we suggest modeling soft tissues as isotropic incompressible hyperelastic materials, using a strain energy function of the form:

$$W = W_1(I_C, II_C) + L(III_C - 1) \qquad \text{leading to a stress tensor} \tag{6.48}$$

$$\mathbf{S}^{(e)}[\mathbf{E}] = \frac{\partial W_1(I_C, II_C)}{\partial \mathbf{E}} + L\frac{\partial III_C}{\partial \mathbf{E}} \tag{6.49}$$

where L is a Lagrange multiplier accounting for the hydrostatic pressure, and to formulate the incremental tangent stiffness matrix as:

$$\mathbf{K}_n^M = \frac{\partial^2 W}{\partial \mathbf{E}_n^2} = \frac{\partial^2 W_1(I_C, II_C)}{\partial \mathbf{E}_n^2} + L_n \frac{\partial^2 I_3}{\partial \mathbf{E}_n^2} \tag{6.50}$$

Suggestions for tendon, skin and passive muscle modeling are presented hereafter.

Tendons and Ligaments. Various constitutive relationships for tendons and ligaments have been presented in Chapt. 4. Due to the parallel orientation of their fibers, most of them have been developed for uniaxial tensile experiments. Nevertheless, a number of 3D models for tendons and ligaments are arising in the form of pseudo-elastic strain energy functions. However, considering that tendons always work in extension, and that the transverse constriction is small compared to the extension, it is possible to use a one-dimensional finite element model, and to assume incompressibility in order to insure the volume deformation on the 3D tendon object. Among the available models, Fung's exponential form remains the most common for soft tissues. We therefore suggest to model tendons and ligaments with a constitutive relation as:

$$T(t) = \int_0^t G(t-\tau)\frac{\partial T^{(e)}[\varepsilon(\tau)]}{\partial \tau}\, d\tau \qquad \text{with} \tag{6.51}$$

$$T^{(e)}(\varepsilon) = a\left(e^{b\varepsilon} - 1\right) \qquad (4.20) \qquad \text{as elastic function,}$$

and with values for the elastic response $T^{(e)}$ and the reduced relaxation function $G(t)$ provided by (Trevisan 83).

Passive Skeletal Muscle. Skeletal muscles have rarely been investigated in the biomechanical literature from a constitutive modeling point of view: as far as we have been in our investigations, we have found only a few uniaxial relations (Fig. 4.23) (Glantz 74, 77), (Fig. 4.24) (Capelo 81). However, a strain energy formulation allowing a 3D non-linear deformation simulation seems more convenient since body visible deformation mainly arises from skeletal muscles. Therefore, we suggest using one of the strain energy functions developed for ventricles, though skeletal muscles may have a slightly different mechanical behavior, in structure, strength, and stiffness. Several constitutive relationships have been presented in Sect. 4.3.3 for cardiac muscle modeling. Among these, we prefer Horowitz's constitutive description (4.95) for the structural properties that the model includes (Horowitz 88c), and because it has already been applied to finite element modeling, as presented in Sect. 6.2.1:

$$W = \sum_k W_k = \sum_k S_k \int_\Omega R_k(\mathbf{u}) w_k^*(\gamma'_{11})\, d\Omega \qquad \text{with} \tag{6.52}$$

$$\mathbf{K}_n^M = S_c \int_{-\frac{\pi}{3}}^{\frac{\pi}{3}} \int_0^{2\pi} \int_0^{\pi} \left[\int_0^{\varepsilon_c} \frac{D_c(x)}{1+2x}\, dx \right] R_c(\alpha, \theta, \phi) \frac{\partial^2 \varepsilon'_n}{\partial \mathbf{E}_n^2} J\, d\phi\, d\theta\, d\alpha + L_n \frac{\partial^2 I_3}{\partial \mathbf{E}_n^2} \tag{6.53}$$

In case this is found far too complex, Humphrey's models described by (4.104) and (4.107) may be convenient substitute suggestions (Humphrey 87–90). For any selected, we suggest Best's data for the QLV formulations, since these have been especially investigated for skeletal muscles (Sect. 4.1.4) (Best 94).

Skin. In contrast to tendons and skeletal muscles, numerous strain energy functions have been proposed for skin modeling. We mention for example Allaire's model which has been applied for in vivo human skin (4.74) (Allaire 77):

$$\begin{cases} W = C_1(I_1 - 3) + C_2(I_2 - 3) + g(I_3) \\ g(I_3) = C_3(I_3 - 1)^2 - (C_1 + 2C_2)\,(I_3 - 1) \end{cases} \tag{6.54}$$

However, considering the reduced thickness of skin with respect to the two other dimensions, it may be more appropriate to consider skin as a membrane, and to model its plane state of stress with a biaxial constitutive relationship such as Tong's exponential biaxial model (4.89) described in (Tong 76):

$$W = f(\alpha, \mathbf{E}) + \frac{c}{2} e^{F(a,\mathbf{E})} \qquad f(\alpha, \mathbf{E}) = \frac{1}{2}\left(\alpha_1 E_{11}^2 + \alpha_2 E_{22}^2 + 2\alpha_4 E_{11} E_{22}\right) \tag{6.55}$$

$$F(a, \gamma, \mathbf{E}) = a_1 E_{11}^2 + a_2 E_{22}^2 + a_3 E_{12}^2 + 2a_4 E_{11} E_{22} + \gamma_1 E_{11}^3 + \gamma_2 E_{22}^3 + \gamma_4 E_{11}^2 E_{22} + \gamma_5 E_{11} E_{22}^2$$

Moreover, in case Horowitz's model is chosen for skeletal muscle modeling, it may be interesting for skin modeling to consider Shoemaker's fibrous QLV model described in (4.120) (Shoemaker 86) as:

$$\mathbf{S} = \mathbf{S}^F + \mathbf{S}^C \quad \text{with} \tag{6.56}$$

$$\mathbf{S}^F = \int_{-\frac{\pi}{2}}^{\frac{\pi}{2}} D(\theta) S_f \frac{\partial E_f}{\partial \mathbf{E}} \, d\theta \quad \text{fibrous stress similar to Lanir's form} \tag{6.57}$$

$$\mathbf{S}^C = \Lambda \int_{-\infty}^{t} g(t-\tau) \frac{\partial \mathbf{E}(\tau)}{\partial \tau} d\tau \quad \text{compliant stress similar to Fung's QLV form} \tag{6.58}$$

$$S_f = \int_{-\infty}^{t} G(t-\tau) \frac{\partial E_f(\tau)}{\partial \tau} d\tau \quad \text{with} \quad G(t) = G_0 g(t) \quad \text{and} \tag{6.59}$$

$$g(t) = \frac{1 + a\int_0^{\infty} f(\tau) e^{-\frac{t}{\tau}} \, d\tau}{1 + a\int_0^{\infty} f(\tau) \, d\tau} \quad \text{with} \quad f(t) = e^{-pt} \quad \text{and} \quad a \quad \text{a constant} \tag{6.60}$$

As both Horowitz's and Shoemaker's models are based on Lanir's structural description, some of the constitutive finite element routines to develop may be used commonly for both muscle and skin tissues simulation with respective data.

6.2.4 Finite Element Meshing

The choice of the element type for finite element meshing procedure depends on the type of analysis to be performed and the expected level of precision. The common finite element software provides extended libraries of element types, depending on the number of geometrical and integration nodes, as well as on their disposition and associated properties. The meshing procedure is therefore closely related to the intrinsic properties of the material and consists in an entire step in the modeling approach. However, some general features may already be outlined here to give some precision on the way soft tissue simulation may be achieved.

Volume Meshing. Our goal is three-dimensional modeling of the skeletal soft-tissues, so volume element meshing seems appropriate. Mainly two kinds of shapes are available for volume elements: tetrahedrons and hexahedrons (Fig. 3.3). Given the complex shapes of muscles, we suggest the use of tetrahedron elements.

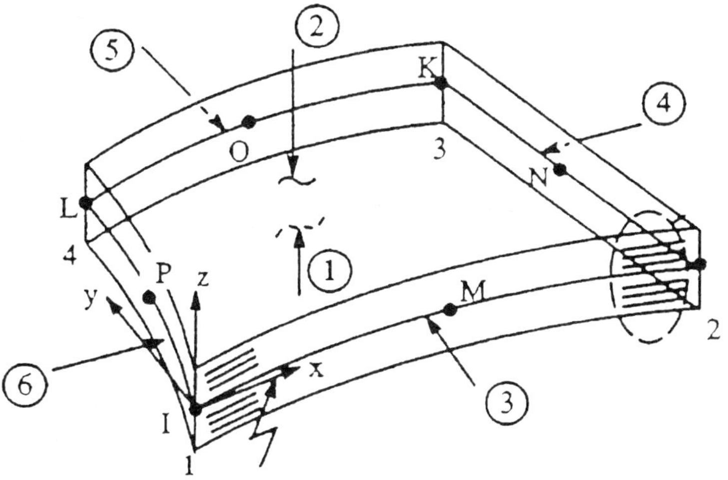

Fig. 6.8. 3D-shell finite element for skin
(reprinted from Ansys User's Manual)

Line Meshing. The deformation of the tendon is strongly oriented in the direction of the fibers. Its volume is also reduced compared to muscle volume. Therefore, as explained previously, it is possible to apply a 1D constitutive relationship with a 1D meshing in the direction of the tension, and to generate volume deformation in assuming incompressibility using a constriction coefficient such as the Poisson ratio ν.

Surface Meshing. The case of skin modeling is more difficult to specify because of the great variability of its structure and properties around the body. Furthermore, constitutive equations for skin usually account for epidermis and dermis only. A simple approach would be to apply the skin constitutive relationship to the three layers considered as a whole. However, as the mechanical properties of the hypodermis are quite different from those of the upper layers, and have significant influence on the appearance and overall behavior of skin, it would be more convenient to distinguish in the model between the constitutive relationships of the upper and subcutaneous layers. This is possible by using two-layered thick composite shell elements (Fig. 6.8) with a different constitutive relationship for each of them. However, as no relationship is available for hypodermis, we suggest using a simple incompressible linear viscoelastic relation or the same relation than the one for epidermis and dermis with user-defined parameters. The thickness of the upper layer may be considered uniform, while a nodal thickness may be specified for the hypodermis layer to account for its thickness variation.

Interconnections Modeling. In real motion, the in-vivo deformation of the different anatomical structures do not occur successively: they happen simultaneously and depend on each other. The deformation of an isolated contracting muscle will not be the same as that of a muscle connected to bones with tendons, and compressed by other deforming muscles. Therefore, the deformation computation must be performed simultaneously on the different muscles involved in the motion. For this purpose, common finite element software usually provides a set of interface elements allowing the resolution of contact dynamics. Mainly two cases appear: *static contacts*, for which nodes on both socks may be matched together, and *sliding contacts* which involve friction and relative displacements of surfaces. In the human anatomical structure, although some may be separated more easily than others, all layers appear more or less tied to their neighbors by a network of infiltrating collagen fibers. Hence, epidermis, dermis, hypodermis, fascia, muscles, tendons and bones are all embedded in a same continuous collagenous network. Therefore, we suggest the use of *static interface* elements for the simultaneous deformation of the different structures.

Depending on the complexity of the analysis, however, a simplification may be made here in considering the insignificance of skin tension on muscle deformation. Skin is much more flexible and lighter than the other soft tissues. It may be assumed that it will just adjust to the deformation of the other stronger soft tissues rather than produce a change in their equilibrium state. It seems therefore possible to compute the deformation of all involved muscles together, and post-process the deformation of skin for each step of the analysis to fit onto the outer surface of all internal objects. While simultaneous calculations use interface elements and require loads specification as input, the constraints for skin deformation analysis would be the specification of the displacements in order to fit onto the deformed envelope of muscles.

6.3 Suggestion for Muscle Contraction Simulation

6.3.1 Equations of Motion

The methodology for applying the finite element method to biomechanical relationships is quite common: whichever the material, the incremental stress-strain relationship is substituted for the stress in the internal component of the virtual works equation (6.1), which may then be solved using geometric discretization and incremental integration. However, this is true as long as the material is passive. Active muscle contraction is such a particular behavior that no model or simulation methodology have found a rule so far. Each model therefore leads to its specific application. This leads us to develop hereafter an approach for contraction simulation based on the geometric description presented in Sect. 5.3.2.

For a proper formulation, it is necessary to refer back to the original formulation of the equation of motion, in considering the contraction phenomenon from a physical point of view. Given the structural and functional constitution of muscle described in Sect. 1.2.1, the active contraction phenomenon appears as arising from multiple contractile mechanisms, microscopically inserted within the passive constituents of the muscle material. Thus, the contraction force appears as internal stresses distributed through the muscle volume, though it rather belongs in nature to the group of the external actions applied onto the passive muscle body. Under these microscopic actions, the muscle deforms in involving its passive elastic and viscoelastic properties. As a result of this description, the equations of motion (2.34) and (2.35) may be rewritten as follows:

Force balance: $$\int_v \rho \mathbf{\Gamma}^v \, dv = \int_v \mathbf{f}^v \, dv + \int_s \mathbf{t}^s \, ds + \int_v \mathbf{f}^c \, dv \qquad (6.61)$$

Moment balance: $$\int_v \rho \mathbf{\Gamma}^v \wedge \mathbf{x} \, dv = \int_v \mathbf{f}^v \wedge \mathbf{x} \, dv + \int_s \mathbf{t}^s \wedge \mathbf{x} \, ds + \int_v \mathbf{f}^c \wedge \mathbf{x} \, dv \qquad (6.62)$$

where $\mathbf{\Gamma}^v$ is the acceleration of point P, $\mathbf{t}^s$ the external stress on ds
$\mathbf{f}^v$ is the external body force, $\mathbf{f}^c$ the contraction force (5.41)

As developed in Sect. 2.1.3, (6.61) leads to the local equation of motion, and (6.62) to the symmetry property of the Cauchy stress tensor $\boldsymbol{\sigma}$:

$$\begin{cases} \rho \mathbf{\Gamma}^v = \mathbf{f}^v + \mathrm{div}(\boldsymbol{\sigma}) + \mathbf{f}^c \\ \boldsymbol{\sigma} = \boldsymbol{\sigma}^T \end{cases} \quad \text{or} \quad \begin{cases} \rho \dfrac{\partial^2 u_i}{\partial t^2} = f_i + \sum_j \dfrac{\partial \sigma_{ij}}{\partial x_j} + f_i^c \\ \sigma_{ij} = \sigma_{ji} \end{cases} \qquad (6.63)$$

Following Sect. 2.1.4, the Eulerian virtual work equation may be developed as:

$$\delta \hat{W}^{acc} = \delta \hat{W}^{ext} + \delta \hat{W}^{int} \qquad (6.64)$$

with $\delta \hat{W}^{acc} = \int_v \delta \hat{\mathbf{u}}^T \rho \mathbf{\Gamma}^v \, dv$ inertial virtual work

$\delta \hat{W}^{int} = -\int_v \mathrm{Tr}(\delta \hat{e}^T \boldsymbol{\sigma}) \, dv = -\int_v \delta \hat{\boldsymbol{e}}^T \boldsymbol{\sigma} \, dv$ internal virtual work

$\delta \hat{W}^{ext} = \int_v \delta \hat{\mathbf{u}}^{f^T} (\mathbf{f}^v + \mathbf{f}^c) \, dv + \int_s \delta \hat{\mathbf{u}}^{t^T} \mathbf{t}^s \mathbf{n} \, ds$ external virtual work

This equation may be converted with respect to the Lagrangian configuration into:

$$\delta\hat{W}^{acc} = \delta\hat{W}^{ext} + \delta\hat{W}^{int} \tag{6.65}$$

with $\delta\hat{W}^{acc} = \int_V \delta\hat{\mathbf{u}}^T \rho \boldsymbol{\Gamma}^V \, dV$ inertial virtual work

$$\delta\hat{W}^{int} = -\int_V \mathrm{Tr}(\delta\mathbf{E}^T\mathbf{S})\, dV = -\int_V \delta\hat{\boldsymbol{E}}^T \boldsymbol{S}\, dV \quad \text{internal virtual work}$$

$$\delta\hat{W}^{ext} = \int_V \delta\hat{\mathbf{u}}^{fT}\left(\mathbf{f}^V + \mathbf{f}^C\right) dV + \int_S \delta\hat{\mathbf{u}}^{tT}\mathbf{t}^S\mathbf{n}\, dS \quad \text{external virtual work}$$

where $\boldsymbol{S}$, $\boldsymbol{E}$ are the *vectorial* forms of the tensors $\mathbf{S}$ and $\mathbf{E}$ as defined in (2.53), and $\mathbf{f}^V$, $\mathbf{f}^C$, $\mathbf{t}^S$ are the Lagrangian equivalents of the Eulerian forces $\mathbf{f}^v$, $\mathbf{f}^c$, $\mathbf{t}^s$.

From (2.2) and (2.6), $\mathbf{f}^C$ and $\mathbf{f}^c$ are related:

$$\mathbf{f}^C = \mathrm{J}\,\mathbf{G}\,\mathbf{f}^c \tag{6.66}$$

6.3.2 Incremental Finite Element Formulation

In the general case, as the problem involves non-linearities, (6.65) must be converted into its incremental form as (3.43) (Sect. 3.2.2):

$$\delta\hat{W}^{acc}_{n+1} = \delta\hat{W}^{1}_{n+1} + \delta\hat{W}^{2}_{n+1} + \delta\hat{W}^{3}_{n+1} + \delta\hat{W}^{ext}_{n+1} \tag{6.67}$$

with $\delta\hat{W}^{acc}_{n+1} = \int_V \delta\hat{\mathbf{u}}^T_{n+1}\rho_n \boldsymbol{\Gamma}^V_{n+1}\, dV$ inertial virtual work

$$\delta\hat{W}^{1}_{n+1} = -\int_V \delta\left(\Delta\hat{\boldsymbol{E}}^L\right)^T\left[\boldsymbol{S}_n + \boldsymbol{S}^R_n\right] dV \quad \text{internal force virtual work}$$

$$\delta\hat{W}^{2}_{n+1} = -\int_V \delta\left(\Delta\hat{\boldsymbol{E}}^L\right)^T \boldsymbol{K}^M_n \Delta\boldsymbol{E}^L\, dV \quad \text{linear strain virtual work}$$

$$\delta\hat{W}^{3}_{n+1} = -\int_V \delta\left(\Delta\hat{\boldsymbol{E}}^{NL}\right)^T\left[\boldsymbol{S}_n + \boldsymbol{S}^R_n\right] dV \quad \text{geometric virtual work}$$

$$\delta\hat{W}^{ext}_{n+1} = \int_V \delta\hat{\mathbf{u}}^{fT}_{n+1}\left(\mathbf{f}^V_{n+1} + \mathbf{f}^C_{n+1}\right) dV + \int_S \delta\hat{\mathbf{u}}^{tT}_{n+1}\mathbf{t}^S_{n+1}\mathbf{N}\, dS \quad \text{external virtual work}$$

Given the semi-dependence of the contraction force $\mathbf{f}^C$ on the local and global muscle deformations, it is here necessary to develop its incremental form as:

$$\mathbf{f}^{C}_{n+1} = \mathbf{f}^{C}_{n} + \Delta\mathbf{f}^{C} \qquad \text{with} \qquad \Delta\mathbf{f}^{C}(P) = \Delta\mathbf{f}^{\omega}(P) + \Delta\mathbf{f}^{M}(P) + \Delta\mathbf{f}^{\delta}(P) \tag{6.68}$$

with $\Delta\mathbf{f}^{\omega}(P) = J_n \Delta\omega(P)\, f^{M}_{n}\left(l^{M}, v^{M}\right) \boldsymbol{\partial}_n(P)$ due to the active strength change

$\Delta\mathbf{f}^{M}(P) = J_n \omega_n(P)\, \Delta f^{M}\left(l^{M}, v^{M}\right) \boldsymbol{\partial}_n(P)$ due to the uniform force change

$\Delta\mathbf{f}^{\delta}(P) = J_n \omega_n(P)\, f^{M}_{n}\left(l^{M}, v^{M}\right) \Delta\boldsymbol{\partial}_n(P)$ due to the fiber direction change

and $\boldsymbol{\partial}_n = \mathbf{G}_n \boldsymbol{\delta}_n$ $\quad\boldsymbol{\partial}$: Lagrangian equivalent of the fiber direction $\boldsymbol{\delta}$ (5.43)

Assuming then a finite element discretization as for (3.47), the incremental finite element virtual work equation becomes:

$$\mathbf{M}_n \ddot{\mathbf{U}}_{n+1} + \left(\mathbf{K}^{L}_{n} + \mathbf{K}^{NL}_{n}\right)\Delta\mathbf{U} = \mathbf{L}_{n+1} - \mathbf{R}_n + \Delta\mathbf{R}^{\omega}_{n} + \Delta\mathbf{R}^{M}_{n} + \Delta\mathbf{R}^{\delta}_{n} \tag{6.69}$$

with $\mathbf{M}_n = \int_V \rho_n \mathbf{H}^T\mathbf{H}\, dV$ as tangent mass matrix

$\mathbf{K}^{L}_{n} = \int_V \boldsymbol{B}^{L^T}_{n} \boldsymbol{K}^{M}_{n} \boldsymbol{B}^{L}_{n}\, dV$ as linear stiffness matrix

$\mathbf{K}^{NL}_{n} = \int_V \boldsymbol{B}^{NL^T}_{n} \left[\boldsymbol{S}_n + \boldsymbol{S}^{R}_{n}\right] \boldsymbol{B}^{NL}_{n}\, dV$ as geometric stiffness matrix

$\mathbf{L}_{n+1} = \int_V \mathbf{H}^{f^T} \mathbf{f}^{V}_{n+1}\, dV + \int_S \mathbf{H}^{t^T} \mathbf{t}^{S}_{n+1} \mathbf{N}\, dS$ as external force vector

$\mathbf{R}_n = \int_V \boldsymbol{B}^{L^T}_{n} \left[\boldsymbol{S}_n + \boldsymbol{S}^{R}_{n}\right] dV - \int_V \mathbf{H}^T \mathbf{f}^{C}_{n}\, dV$ as internal force vector

$\Delta\mathbf{R}^{\omega} = \int_V \mathbf{H}^T \Delta\mathbf{f}^{\omega}\left(\mathbf{U}, \dot{\mathbf{U}}\right) dV$ as active strength increment vector

$\Delta\mathbf{R}^{M} = \int_V \mathbf{H}^T \Delta\mathbf{f}^{M}\left(\mathbf{U}, \dot{\mathbf{U}}\right) dV$ as uniform force increment vector

$\Delta\mathbf{R}^{\delta} = \int_V \mathbf{H}^T \Delta\mathbf{f}^{\delta}\left(\mathbf{U}, \dot{\mathbf{U}}\right) dV$ as direction force increment vector

Taking into account the geometrical nature of the contraction force model proposed in Sect. 5.3, profit may be done of the finite element approach in applying on the mesh nodes contraction forces obtained from a respective finite element discretization of the contraction force distribution density. Each node of the finite element mesh may be assigned its associated fiber direction $\boldsymbol{\delta}$ and active strength ω, calculated from its coordinates. As a result, the fiber direction $\boldsymbol{\partial}$, and the active strength ω and uniform contraction force f^M, of any point of the muscle, may be interpolated between those of the surrounding nodes, using relations:

$$\boldsymbol{\partial}_n = \mathbf{D}^{\delta}\boldsymbol{\delta}_n \qquad \omega_n = \omega_n^T = \boldsymbol{\Omega}\,\mathbf{w}_n \qquad f_n^M = f_n^{M^T} = \dot{\mathbf{U}}_n^T\,\mathbf{D}^{M^T}\mathbf{H}^M\,\mathbf{U}_n \tag{6.70}$$

where
- $\boldsymbol{\delta}_n$ is the nodal fiber direction vector at step n
- $\mathbf{D}^{\delta}$ is the tensorial shape function for the fiber direction function $\boldsymbol{\delta}$
- $\mathbf{w}_n$ is the nodal fiber active strength vector at step n
- $\boldsymbol{\Omega}$ is the tensorial shape function for the active strength
- $\mathbf{U}_n$, $\dot{\mathbf{U}}_n$ are the nodal displacement, velocity vectors at step n
- $\mathbf{H}^M$, $\mathbf{D}^M$ are the tensorial shape functions for the uniform force

Using (6.70), the contraction force and associated increment vectors become:

$$\mathbf{f}_n^C = -J_n\left(\dot{\mathbf{U}}_n^T\mathbf{D}^{M^T}\mathbf{H}^M\mathbf{U}_n\right)\mathbf{w}_n^T\boldsymbol{\Omega}^T\mathbf{D}^{\delta}\boldsymbol{\delta}_n \tag{6.71}$$

$$\Delta\mathbf{f}^C = \Delta\mathbf{f}^{\omega} + \Delta\mathbf{f}^M + \Delta\mathbf{f}^{\delta} \qquad \text{with} \tag{6.72}$$

$$\Delta\mathbf{f}^{\omega} = -J_n\,\mathbf{D}^{\delta}\boldsymbol{\delta}_n\left(\dot{\mathbf{U}}_n^T\mathbf{D}^{M^T}\mathbf{H}^M\mathbf{U}_n\right)\boldsymbol{\Omega}\Delta\mathbf{w} \qquad \Delta\mathbf{f}^{\delta} = -J_n\left(\dot{\mathbf{U}}_n^T\mathbf{D}^{M^T}\mathbf{H}^M\mathbf{U}_n\right)\mathbf{w}_n^T\boldsymbol{\Omega}^T\mathbf{D}^{\delta}\Delta\boldsymbol{\delta}$$

$$\Delta\mathbf{f}^M = -J_n\left(\mathbf{w}_n^T\boldsymbol{\Omega}^T\mathbf{D}^{\delta}\boldsymbol{\delta}_n\right)\dot{\mathbf{U}}_n^T\mathbf{D}^{M^T}\mathbf{H}^M\Delta\mathbf{U} - J_n\left(\mathbf{w}_n^T\boldsymbol{\Omega}^T\mathbf{D}^{\delta}\boldsymbol{\delta}_n\right)\mathbf{U}_n^T\mathbf{H}^{M^T}\mathbf{D}^M\Delta\dot{\mathbf{U}}$$

As a result, the incremental finite element equation dynamics (6.69) becomes:

$$\mathbf{M}_n\ddot{\mathbf{U}}_{n+1} + \mathbf{D}^C\Delta\dot{\mathbf{U}} + \left(\mathbf{K}_n^L + \mathbf{K}_n^{NL} + \mathbf{K}_n^C\right)\Delta\mathbf{U} - \Delta\mathbf{R}^{\delta} = \mathbf{L}_{n+1} + \Delta\mathbf{R}^{\omega} - \mathbf{R}_n \tag{6.73}$$

$$\mathbf{D}_n^C = \int_V J_n\mathbf{H}^T\left(\mathbf{w}_n^T\boldsymbol{\Omega}^T\mathbf{D}^{\delta}\boldsymbol{\delta}_n\right)\mathbf{U}_n^T\mathbf{H}^{M^T}\mathbf{D}^M dV \quad \text{force-velocity tangent damping matrix}$$

$$\mathbf{K}_n^C = \int_V J_n\mathbf{H}^T\left(\mathbf{w}_n^T\boldsymbol{\Omega}^T\mathbf{D}^{\delta}\boldsymbol{\delta}_n\right)\dot{\mathbf{U}}_n^T\mathbf{D}^{M^T}\mathbf{H}^M\, dV \quad \text{force-length tangent stiffness matrix}$$

$$\Delta\mathbf{R}^{\delta} = \int_V \mathbf{H}^T\Delta\mathbf{f}^{\delta}(\Delta\boldsymbol{\delta})\, dV \qquad \Delta\mathbf{R}^{\omega} = \int_V \mathbf{H}^T\Delta\mathbf{f}^{\omega}(\Delta\mathbf{w})\, dV$$

In order to allow the resolution of (6.73) using common incremental resolution methods, the direction force increment vector $\Delta\mathbf{R}^{\delta}$, and therefore the change in fiber direction, must be assumed negligible, unless a tensorial relationship may be established between $\Delta\boldsymbol{6}$ and $\Delta\mathbf{U}$. The fiber direction must however be updated at the end of each step in order to take into account the fiber deformation in the other active increments. Besides of this, the same shape function $\mathbf{H}$, used for the geometric interpolation may be used, instead of $\boldsymbol{\Omega}$, for interpolating the active strength distributions ω, in order to reduce the amount of computation.

As a result, the contraction process may be simulated following the way it occurs in a real muscle: a change of activation $\Delta\mathbf{w}$ produces a change of contraction force $\Delta\mathbf{f}^{C}$, which exerts on passive muscle tissue along distributed fiber directions, and which results in the deformation. Simultaneously, the deformation produces a change in the internal passive stress, as well as a change in the uniform contraction force f^{M} that the sarcomeres can develop. As a result, the system deforms so that the passive stress equilibrates the contraction force and the external loads. As opposed to the simulation of passive behavior, though increments of external load may still be added at each step, the real input is the change of activation by means of the active strength increment vector $\Delta\mathbf{w}$. This thus leaves to the user the responsibility of learning how to activate the muscle model for exerting a load or controlling its contraction, in the same way humans learn how to control their force and to refine their gestures.

Conclusion

Following the real deformation phenomenon, muscle contraction may be simulated step-by-step within an incremental process generating muscle shortening, tendon lengthening, skeletal motion, and skin deformation. As explained in the introduction, the exact scheme of the simulation depends on the practical aspects of the implementation and the requirements of the finite element code applied. The approach presented here for muscle contraction modeling and simulation has not been applied yet, and is proposed here as basis for future developments. The effective selection, implementation, and simulation of the biomechanical models for muscle, tendon, and skin, using commercial finite element software, may be found in the works of our CHARM partners, the team directed by Professors Martins and Pires in the Civil Engineering Department (DECcivil) in the Superior Technical Institute of Lisbon (IST) (Appendix).

References

Allaire 77 P.E. Allaire, J.G. Thacker, R.F. Edlich, G.J. Rodenheaver, M.T. Edgerton (1977), Finite deformation theory for in-vivo human skin, *J. Bioeng.*, 1, 239–249

Best 94 T.M. Best, J. McElhaney, W.E. Garrett Jr, B.S. Myers (1994), Characterization of the passive responses of live skeletal muscle using the quasi-linear theory of viscoelasticity, *J .Biomechanics*, 27, 413–419

Bro-Nielsen 96 M. Bro-Nielsen, S. Cotin (1996), Real-time volumetric deformable models for surgery simulation using finite elements and condensation, *Proc. Eurographics '96*, *Computer Graphics Forum*, 15, 3, C-57–C-66

Capelo 81 A. Capelo, V. Comincioli, R. Minelli, C. Poggesi, C. Reggiani, L. Ricciardi (1981), Study and parameters identification of a rheological model for excised quiescent cardiac muscle, *J. Biomechanics*, 14, 1–11

Celniker 91 G. Celniker, D. Gossard (1991), Deformable curve and surface finite-elements for free-form shape design, *Proc. Siggraph '91*, *Computer Graphics*, 25, 257–266

Chen 92 D.T. Chen, D. Zeltzer (1992), Pump it up: computer animation of a biomechanically based model of muscle using the finite element method, *Proc. Siggraph '92*, *Computer Graphics*, 26, 89–98

Christie 80 G.W. Christie, I.C. Medland (1980), A static finite strain, finite element stress analysis of bioprosthetic heart valves, *Proc. Int. Conf. on Finite Elements in Biomechanics*, Feb. 18–20, University of Arizona

Christie 82 G.W. Christie, I.C. Medland (1982), A non-linear finite element stress analysis of bioprosthetic heart valves, in *Finite Elements in Biomechanics*, ed. by R.H. Gallagher, B.R. Simon, P.C. Johnson, J.F. Gross. Chichester, UK: John Wiley & Sons

Deng 88 X.Q. Deng (1988), A finite element analysis of surgery of the human facial tissue, Ph.D. thesis, Columbia University, New York

Diamant 72 J. Diamant, A. Keller, E. Baer, M. Lii, R.G.C. Arridge (1972), Collagen, ultrastructure and its relations to mechanical properties as a function of ageing, *Proc. R. Soc. London,* B. 180, 293–315

Essa 93 I. Essa, S. Sclaroff, A. Pentland (1993), Physically based modeling for graphics and vision, in *Directions in Geometric Computing*, ed. by Ralph Martin. Winchester, UK: Information Geometers

Feit 79 T.S. Feit (1979), Diastolic pressure-volume relations and distribution of pressure and fiber extension across the wall of a model left ventricle, *Biophys. J.*, 28, 143–166

Fung 72 Y.C. Fung (1972), Stress-strain history relations of soft tissues in simple elongation, in *Biomechanics: Its Foundations and Objectives*, ed. by Y.C. Fung, N. Perrone, M. Anliker. Englewood Cliffs, NJ: Prentice Hall

Fung 93 Y.C. Fung (1993), Bioviscoelastic solids, in *Biomechanics: Mechanical Properties of Living Tissues.* Berlin Springer-Verlag

Galford 70 J.E. Galford, J.H. McElhaney (1970), A viscoelastic study of scalp, brain and dura, *J. Biomechanics*, 3, 211–221

Gascuel 91 M. P. Gascuel, A. Verroust, C. Puech (1991), A modeling system for complex deformable bodies suited to animation and collision processing, *J. Visual. Comp. Anim.*, 2

Glantz 74 S.A. Glantz (1974), A constitutive equation for the passive properties of muscle, *J. Biomechanics* , 7, 137–145

Glantz 77 S.A. Glantz (1977), A three-element description for muscle with viscoelastic passive elements, *J. Biomechanics* , 10, 5–20

Gourret 89 J.P. Gourret, N. Magnenat-Thalmann, D. Thalmann (1989), Simulation of object and human skin deformations in a grasping task, *Proc. Siggraph '89, Computer Graphics*, 23, 21–30

Gourret 91 J.P. Gourret, N. Magnenat-Thalmann, D. Thalmann (1991), Modeling of contact deformations between a synthetic human and his environment, *Computer Aided Design*, 23, 7, 514–527

Haut 69 R.C Haut, R.W. Little (1969), The rheological properties of canine anterior cruciate ligaments, *J. Biomechanics*, 2, 289–298

Hirsch 68 C. Hirsch, L. Sonnerup (1968), Macroscopic rheology in collagen material, *J. Biomechanics*, 1, 13–18

Horowitz 88a A. Horowitz, I. Sheinman, Y. Lanir, M. Perl, S. Sideman (1988), Non-linear incompressible finite element for simulating loading and unloading of cardiac tissue – Part I: Two-dimensional formulation for thin myocardial strips, *J. Biomech. Engng.*, 110, 57–62

Horowitz 88b A. Horowitz, I. Sheinman, Y. Lanir (1988), Non-linear incompressible finite element for simulating loading and unloading of cardiac tissue – Part II: Three-dimensional formulation for thick ventricular wall segments, *J. Biomech. Engng.*, 110, 62–68

Horowitz 88c A. Horowitz, I. Sheinman, Y. Lanir, M. Perl, R.K. Strumpf (1988), Structural three dimensional constitutive law for the passive myocardium, *J. Biomech. Engng.*, 110, 200–207

Huang 90 X. Huang, M.M. Black, I.C. Howard, E.A. Patterson (1990), A two-dimensional finite element analysis of a bioprosthetic heart valve, *J. Biomechanics*, 23, 753–762

Humphrey 87 J.D. Humphrey, F.C.P. Yin (1987), On constitutive relations and finite deformations of passive cardiac tissue – Part I: A pseudo-strain energy function, *J. Biomech. Engng.*, 109, 298–304

Humphrey 90 J.D. Humphrey, R.K. Strumpf, F.C.P. Yin (1990), Determination of a constitutive relation for passive myocardium – Parts I and II, *J. Biomech. Engng.*, 112, 333–346

Huygue 91 J. M. Huygue, D.H van Campen, T. Arts, R.M. Heethaars (1991), A two-phase finite element model of the diastolic left ventricle, *J. Biomechanics*, 24, 527–538

Janz 73 R.F. Janz, A.F. Grimm (1973), Deformation of the diastolic left ventricle – Part I: Non-linear elastic effects, *Biophys. J.*, 13, 689–704

Janz 74 R.F. Janz, R.K. Bruce, T.F. Moriarty (1973), Deformation of the diastolic left ventricle – Part II: Non-linear geometric effects, *J. Biomechanics*, 7, 509–516

Karakaplan 80 A.D. Karakaplan, M.P. Bienek, R. Skalak (1980), A mathematical model of lung parenchyma, *J. Biomech. Engng.*, 102, 124–136

Klosner 69 J.M. Klosner, A. Segal (1969), Mechanical characterization of a natural rubber, *PIBAL Rep.* 69–42, Polytechnic Institute of Brooklyn, New York

Koch 96 R.M. Koch, M.H. Gross (1996), Simulating facial surgery using finite element models, *Proc. Siggraph '96, Computer Graphics*, 30, 421–428

Kwan 89 M.K. Kwan, S L-Y. Woo (1989), A structural model to describe the non-linear stress-strain behavior for parallel-fibered collagenous tissues, *J. Biomech. Engng.*, 111, 361–363

Lanir 83 Y. Lanir (1983), Constitutive equations for fibrous connective tissues, *J. Biomechanics*, 16, 1–12

Larrabee 86 W.F. Larrabee Jr, J.A. Galt (1986), A finite element model of skin deformation – Part III: The finite element model, *Laryngoscope '96*, 413–419

Lee 83 G.C. Lee, N.T. Tseng, Y.M. Yuan (1983), Finite element modeling of lungs including interlobar fissures and the heart cavity, *J. Biomechanics*, 16, 9, 679–690

Liu 78 J.T. Liu, G.C. Lee (1978), Static finite deformation analysis of the lung, *J. Engng. Mech. Div.*, ASCE, 104, (EMI) 225–238

Needleman 83 A. Needleman, S.A. Rabinowitz, D.K. Bogen, T.A. McMahon (1983), A finite element model of the infarcted left ventricle, *J. Biomechanics*, 16, 45–58

Pao 78 Y.C. Pao, P.A. Chevalier, J.R. Rodarte, L.D. Harris (1978), Finite element analysis of the strain variations in excised lobe of canine lung, *J. Biomechanics*, 11, 91–100

Pieper 92 S.D. Pieper (1992), CAPS: Computer-Aided Plastic Surgery, Ph. D. Dissertation, MIT

Platt 81 S. Platt, N.I. Badler (1981), Animating facial expressions, *Proc. Siggraph '81, Computer Graphics*, 15, 3, 245–252

Sanjeevi 82 R. Sanjeevi, N. Somanathan, D. Ramaswamy (1982), A viscoelastic model for collagen fibres, *J. Biomechanics*, 15, 181–183

Sederberg 86 T.W. Sederberg, S. Parry (1986), Free-form deformations of solid geometric models, *Proc. Siggraph '86, Computer Graphics*, 20, 151–160

Shoemaker 86 P.A. Shoemaker, D. Scheider, M.C. Lee, Y.C. Fung (1986), A constitutive model for two-dimensional soft tissues and its application to experimental data, *J. Biomechanics*, 19, 6, 695–702

Taber 91a L.A. Taber (1991), On a non-linear theory for muscle shells – Part I: Theoretical development, *J. Biomech. Engng.*, 113, 56–62

Taber 91b L.A. Taber (1991), On a non-linear theory for muscle shells – Part II: Application to the beating of the left ventricle, *J. Biomech. Engng.*, 113, 63–71

Terzopoulos 87 D. Terzopoulos, J. Platt, A. Barr, K. Fleischer (1987), Elastically deformable models, *Proc. Siggraph '87, Computer Graphics*, 21, 205–214

Terzopoulos 88 D. Terzopoulos, K. Fleischer (1988), Modeling inelastic deformation: viscoelasticity, plasticity, fracture, *Proc. Siggraph '88, Computer Graphics*, 22, 269–278

Terzopoulos 90 D. Terzopoulos, K. Waters (1990), Physically-based facial modeling, analysis and animation, *J. Visual. Comp. Anim.*, 1, 73–80

Terzopoulos 91 D. Terzopoulos, K. Waters (1991), Techniques for realistic facial modeling and animation, *Proc. Computer Animation'91*, ed. by N. Magnenat-Thalmann, D. Thalmann. Tokyo: Springer-Verlag

Tong 76 P. Tong, Y.C. Fung (1976), The stress-strain relationship for the skin, *J. Biomechanics*, 9, 649–657

Trevisan 83 Trevisan (1983), Etude des propriétés rhéologiques des tissus biologiques. Application au comportement mécanique des ligaments naturels et artificiels, Ph. D. Thesis, Paris XII

Vawter 80 D.L. Vawter (1980), A finite element model for macroscopic deformation of the lung, in *Finite Elements in Biomechanics*, ed. by R.H. Gallagher, B.R. Simon, P.C. Johnson, J.F. Gross. Chichester, UK: John Wiley & Sons

Waters 87 K. Waters (1987), A muscle model for animating three-dimensional facial expressions, *Proc. Siggraph '87, Computer Graphics*, 21, 17–24

Wertheim 47 M.G. Wertheim (1847), Mémoire sur l'élasticité et la cohésion des principaux tissus du corps humain, *Annls. Chim. Phys.*, 21, 385–414

Wu 95 Y. Wu, N. Magnenat-Thalmann, D. Thalmann (1995), A dynamic wrinkle model in facial animation and skin ageing, *J. Visual. Comp. Anim.*, 6, 195–205

Appendix

ESPRIT 9036 Basic Research Project CHARM

<http://ligwww.epfl.ch/~maurel/CHARM/>

The CHARM project is a 3-year basic research project initiated in November 1993 by the European Commission under the ESPRIT program. The European team has been coordinated by *Universitat de les Illes Balears* (UIB) and comprises as partners: *École Polytechnique Fédérale de Lausanne* (EPFL), *Université de Genève* (UG), *Universität Karlsruhe* (UK), *Instituto Superior Técnico de Lisboa* (IST), *École des Mines de Nantes* (EMN), and as subcontractors *IRISA* and *Hôpital Cantonal Universitaire de Genève*. The acronym CHARM stands for *A Comprehensive Human Animation Resource Model*. The objective of CHARM is to develop a 3D biomechanical human model allowing the realistic dynamic simulation of the complex human musculoskeletal system, including the finite element simulation of soft tissue deformation and muscular contraction. After several months of work, the target of developing a comprehensive model has been perceived as too ambitious for a 3-year project. Initial work has thus concentrated on the shoulder-arm complex, which is commonly considered as one of the most complex articulations in the human body.

3D Topological Reconstruction. The three-dimensional reconstruction of the human upper limb has been achieved in UG, using the imaging data of the Visible Human Project initiated by the U.S. National Library of Medicine. Segmentation has been performed by anatomists using the *labeling tool*, which has been developed to allow the semi-automatic segmentation of the anatomical structures on different modalities. Another tool, the *topological modeler*, has been developed to allow the interactive definition of the mechanical attributes and topological inter-relationships of the different anatomical structures.

Biomechanical Modeling. In order to perform the dynamic simulation, mechanical properties must be assigned to the various anatomical components of the human model. The topological modeler developed in UG has thus been used by EPFL to construct the underlying biomechanical model with the assistance of orthopedic specialists. The model defines the joint kinematics, the topology of muscle actions, and the rigid body mechanical properties required for a dynamic simulation. The biomechanical literature has also been investigated for reviewing the available models of soft tissues required for the finite element analysis.

High Level Motion Control. The motion simulation of the human arm model defined in EPFL has been achieved by EMN by means of a numerical solver, including the equation dynamics automatically generated by the NMECAM system developed in IRISA. Various control strategies, including linear, proportional-derivative, and constraint-based strategies, have been developed and applied to the motion control of the human arm complex. A high-level interface layer has finally been developed in order to allow the triggering of movements with natural language expressions containing qualitative attributes.

Finite Element Analysis. Due to the redundancies of the musculoskeletal system, optimization analyses has been performed by IST, using muscles topology, to compute, for specified movements, the respective muscle force contributions. The resulting contraction forces have then been used as an input to the ABAQUS finite element code to simulate muscle contraction and soft tissues deformation. From the biomechanical review provided by EPFL, various constitutive relationships have been implemented for muscles, tendons, and skin.

Photorealistic Rendering. Using skeletal motion data and finite element deformation sequences, the photorealistic rendering of the simulation has been achieved by UK. For this purpose, synthetic procedural textures have been developed following spatial spectra texture resynthesis approaches, based on spatial spectra and feature superimposition. A new raytracer has been developed, which can efficiently handle large amounts of surface with high texturing quality, and produce texture deformations for soft tissue rendering.

Developments for Validation. Real test sequences have finally been generated in UIB for the comparative validation of the model. The aim is to base motion validation on reconstructed fluoroscopic sequences of movements and surface deformation validation on real video sequences. For this purpose, an interface allowing multimodal matching has been developed and used to compare the synthetically generated movements to the real motion.

The developments achieved in CHARM constitute an initial step for further advances in research. Anatomists have found the VHD model to be a gold standard for structure identification and labeling. The models developed could be of use for medical education, as well as clinical applications like surgical simulation. Orthopedicians and sports scientists consider our developments as offering unprecedented insights into the complex kinematics of the articulations and into the unknown neuro-muscular control strategies. Finally, from our point of view, as members of the computer graphics research community, CHARM constitutes an outstanding contribution towards the realistic physically-based modeling and animation of the human body.

Related Publications (up to July 97)

Ballester 94	C. Ballester, V. Caselles, M. Gonzalez (1994), Affine invariant segmentation by variational methods, *Proc. 9th Cong. Rec. Form. Int. Art. '94,* Paris 11–14 Janvier 1994 (AFCET)
Beylot 96	P. Beylot, P. Gingins, P. Kalra, N. Magnenat-Thalmann, W. Maurel, D. Thalman, J. Fasel (1996), 3D interactive topological modeling using visible human dataset, *Proc. Eurographics '96, Computer Graphics Forum*, 15, 3, C-33–C-44
Caselles 95	V Caselles, R Kimmel, G Sapiro (1995), Geodesic active contours, *Proc. Int. Conf. Comp. Vis. '95*, 694–699
Gingins 96a	P. Gingins, P. Beylot, P. Kalra, N. Magnenat-Thalmann, W. Maurel, D. Thalmann, J. Fasel (1996), Modeling using the Visible Human Dataset, *Proc. Medical Informatics Europe*, IOS Press, 739–743
Gingins 96b	P. Gingins, P. Kalra, P. Beylot, N. Magnenat-Thalmann, J. Fasel (1996), Using VHD to build a comprehensive human model, *The Visible Human Project Conference*, Oct.7-8, Bethesda, MD, 1996
Kalra 95	P. Kalra, P. Beylot, P. Gingins, N. Magnenat-Thalmann, P. Volino, P. Hoffmeyer, J. Fasel, F. Terrier (1995), Topological modeling of human anatomy using medical data, *Computer Animation* '95, 172–180
Martins 97	J.A.C. Martins, E.B. Pires, L.R. Salvado, P.B. Dinis (1997), A numerical model of the passive and active behavior of skeletal muscles, to appear in *Computer Methods in Applied Mechanics and Engineering*
Mateus 95	C.M. Mateus, J.A.C. Martins, J.C. Maia, E.B. Pires (1995), Static and dynamic optimization in redundant biomechanical systems, *Optimization'95* , Braga, Portugal
Maurel 96	W. Maurel, D. Thalmann, P. Hoffmeyer, P. Beylot, P. Gingins, P. Kalra, N. Magnenat-Thalmann (1996), A biomechanical musculoskeletal model of human upper limb for dynamic simulation, in *Proc. 7th Eurographics Int. Works. Comp. Anim. Sim. '96*, ed. by R. Boulic, G. Hegron. Vienna: Springer-Verlag
Pires 97	J.A.C. Martins, E.B. Pires, L.R. Salvado, P.B. Dinis (1997), A finite element model for the behavior of skeletal muscles, invited lecture, *International Conference on Applied Analysis*, Lisbon

Esprit Basic Research Series

J. W. Lloyd (Ed.): **Computational Logic.** Symposium Proceedings, Brussels, November 1990. XI, 211 pages. 1990

E. Klein, F. Veltman (Eds.): **Natural Language and Speech.** Symposium Proceedings, Brussels, November 1991. VIII, 192 pages. 1991

G. Gambosi, M. Scholl, H.-W. Six (Eds.): **Geographic Database Management Systems.** Workshop Proceedings, Capri, May 1991. XII, 320 pages. 1992

R. Kassing (Ed.): **Scanning Microscopy.** Symposium Proceedings, Wetzlar, October 1990. X, 207 pages. 1992

G. A. Orban, H.-H. Nagel (Eds.): **Artificial and Biological Vision Systems.** XII, 389 pages. 1992

S. D. Smith, R. F. Neale (Eds.): **Optical Information Technology.** State-of-the-Art Report. XIV, 369 pages. 1993

Ph. Lalanne, P. Chavel (Eds.): **Perspectives for Parallel Optical Interconnects.** XIV, 417 pages. 1993

D. Vernon (Ed.): **Computer Vision: Craft, Engineering, and Science.** Workshop Proceedings, Killarney, September 1991. XIII, 96 pages. 1994

E. Montseny, J. Frau (Eds.): **Computer Vision: Specialized Processors for Real-Time Image Analysis.** Workshop Proceedings, Barcelona, September 1991. X, 216 pages. 1994

J. L. Crowley, H. I. Christensen (Eds.): **Vision as Process.** Basic Research on Computer Vision Systems. VIII, 432 pages. 1995

B. Randell, J.-C. Laprie, H. Kopetz, B. Littlewood (Eds.): **Predictably Dependable Computing Systems.** XIX, 588 pages. 1995

F. Baccelli, A. Jean-Marie, I. Mitrani (Eds.): **Quantitative Methods in Parallel Systems.** XVIII, 298 pages. 1995

J. F. McGilp, D. Weaire, C. H. Patterson (Eds.): **Epioptics.** Linear and Nonlinear Optical Spectroscopy of Surfaces and Interfaces. XII, 230 pages. 1995

W. Maurel, Y. Wu, N. Magnenat Thalmann, D. Thalmann: **Biomechanical Models for Soft Tissue Simulation.** XVI, 173 pages. 1998